Satellite Farming

Latief Ahmad • Syed Sheraz Mahdi

Satellite Farming

An Information and Technology Based Agriculture

Springer

Latief Ahmad
Division of Agronomy
Sher-e-Kashmir University of Agricultural
Sciences and Technology of Kashmir
Srinagar, Jammu and Kashmir, India

Syed Sheraz Mahdi
Division of Agronomy
Sher-e-Kashmir University of Agricultural
Sciences and Technology of Kashmir
Srinagar, Jammu and Kashmir, India

ISBN 978-3-030-03447-4 ISBN 978-3-030-03448-1 (eBook)
https://doi.org/10.1007/978-3-030-03448-1

Library of Congress Control Number: 2018962553

This Springer imprint is published by the registered company Springer Nature Switzerland AG
The registered company address is: Gewerbestrasse 11, 6330 Cham, Switzerland

Foreword

Sher-e-Kashmir
University of Agricultural Sciences & Technology of Kashmir
www.skuastkashmir.ac.in

Prof. (Dr.) Nazeer Ahmed
FISVS,FHSI,FISHRD & FCHAI
Vice-Chancellor

The day-to-day challenges of resources and environmental degradation have re-necessitated for adoption of innovative farming that increases the resource efficiency while protecting the natural resources. Satellite farming is becoming ever more relevant as the agricultural industry struggles to come to terms with the environment, economics, traceability, vehicle guidance, and crop management. While some benefits have proved elusive, others contribute positively to today's agriculture. Research continues to be necessary and needs to be reported and disseminated to a wide audience.

Satellite farming: An information and technology based agriculture focuses on principles and applications of site-specific crop management, providing information on rapidly evolving agricultural technologies like precision farming which is an agricultural management system uses global navigation satellite systems, geographic information systems , remote sensing, and data management systems for optimizing the use of nutrients, water, seed, pesticides, and energy in heterogeneous field situations. The authors of the book have tried to provide extensive information on the state of the art of research on precision crop protection and recent developments in site-specific application technologies for the management of weeds, arthropod pests, pathogens, and nematodes. It gives the reader an up-to-date and in-depth review of both basic and applied research developments and focuses specifically on PA technologies and discusses the components, tools, sampling, tillage,

VRT, laser leveling, site-specific nutrient management, water management, pest and disease management, recent advances in PA, yield monitoring and mapping, crop simulation models, and accurate record keeping. It also discusses the impact of precision techniques on soil fertility, nutrient dynamics, and crop productivity, and highlights the application of GPS techniques to regulate fertilizer supply based on soil nutrient distribution and yield goals set by the farmers.

This book on Satellite farming has been prepared to enhance the understanding of UG and PG students which covers all the components as per the syllabus framed by ICAR. I congratulate the authors and I am very hopeful that this book will adequately meet the needs of the students of agriculture and its allied sciences.

20-09-2018
Srinagar, India

Nazeer Ahmed

Preface

Rapid socioeconomic changes in some developing countries, including India, are creating new scopes for application of satellite farming. It helps in managing crop production inputs in an environmentally friendly way. By using site-specific knowledge, satellite farming can target rates of fertilizer, seed, and chemicals for soil and other conditions. This book is a comprehensive source of information on the present state of precision agriculture (PA) and its principles, which are important in crop production while using PA technologies. It also discusses about different components, tools, and techniques used in the implementation of PA. The reader can also find a description of types and importance of precision tillage and working of different variable rate technologies and their application. It also includes technical description of laser land leveler, its components, and the procedure of laser-guided land leveling. Site-specific nutrient management (SSNM) has been elaborated which includes the present approach and constraints in the adoption of SSNM particularly in India. The book also includes the methods of precision water, pests, and weed management. Some of the new concepts have been discussed which deals with techniques used for yield monitoring, soil sampling, and mapping in PA. The book also focuses on feasibility of precision farming and the problems faced during adoption of precision farming techniques and its potential in India. The reader can find a description of different decision support tools like crop simulation models, their types, and application in PA. At the end of the book, there is an index/glossary to provide an accessible dictionary of assessment-related terms for use by a nontechnical, lay audience.

The book is designed to give a comprehensive idea to the undergraduate and postgraduate students of Agricultural Universities on basic principles and technologies, crop-specific applications, integrative strategies, economic and environmental concerns, and future trends in precision agriculture. This book contains 13 chapters on different aspects of the subject. Each chapter is briefed with immense wealth of information of practical nature.

The authors are highly grateful to Prof. Sheikh Bilal Ahmad, Dean, Faculty of Agriculture, and Prof. Farooq A. Zaki, Dean, Faculty of Horticulture of SKUAST-K, for their able guidance and encouragement for preparation of this book.

The authors are also grateful to Prof. Raihana Habib Kanth, Head, Division of Agronomy, for her valuable suggestions during writing of the script.

Srinagar, India
September 2018

Latief Ahmad
S. Sheraz Mahdi

Contents

List of Abbreviations

CTF	Controlled Traffic Farming
CF	Compact Flash
DGPS	Differential Global Positioning System
DSMs	Digital Soil Maps
DSS	Decision Support Systems
DTM	Digital Terrain Model
EC	Electrical Conductivity
GCM	Global Circulation Model
GDP	Gross Domestic Product
GIS	Geographical Information Systems
GNSS	Global Navigation Satellite System
GPS	Global Positioning System
IoT	Internet of Things
IPM	Integrated Pest Management
IPNI	International Plant Nutrition Institute
IRRI	International Rice Research Institute
IVR	Interactive Voice Response
IWM	Integrated Weed Management
LCC	Leaf Color Chart
LIDAR	Light Detection and Ranging
MIT	Massachusetts Institute of Technology
MOA	Modes of Action
MSNC	Modulated Spraying-Nozzle Control
NDVI	Normalized Difference Vegetation Index
NRCS	Natural Resources Conservation Service
NSS	National Social Service
PA	Precision Agriculture
PCMCIA	Personal Computer Memory Card International Association
PF	Precision Farming
PLF	Precision Livestock Farming
PV	Precision Viticulture
SPAD	Soil Plant Analytical Development

SSCM	Site-Specific Crop Management
SSNM	Site-Specific Nutrient Management
UAV	Unmanned Aerial Vehicle
VRA	Variable-Rate Application
VRT	Variable-Rate Technology

List of Figures

List of Tables

Chapter 1
Introduction to Precision Agriculture

Abstract The introductory chapter provides information about basics of accuracy, precision, and precision agriculture. The chapter gives an insight of what precision and accuracy are in measurement, their differences, and use. Precision agriculture is a now a term utilized all through farming frameworks around the world. But what do we mean by "precision agriculture"? This chapter offers a background to the evolution of precision agriculture, the principle philosophy and goals of a precision agriculture management strategy, and some of the steps required to adopt precision agriculture in cropping systems.

Keywords Accuracy · Precision · Precision agriculture · History · Objectives · Variability · Smart farming technologies · SSCM

1.1 Accuracy and Precision

Accuracy refers to the closeness of a measured value to a standard or known value. For example, if in lab you obtain a weight measurement of 3.2 kg for a given substance, but the actual or known weight is 10 kg, then your measurement is not accurate. In this case, your measurement is not close to the known value. Accuracy is how close.

Precision refers to the closeness of two or more measurements to each other. Using the example above, if you weigh a given substance five times, and get 3.2 kg each time, then your measurement is very precise. Precision is independent of accuracy. You can be very precise but inaccurate, as described above. You can also be accurate but imprecise.

For example, if on average, your measurements for a given substance are close to the known value, but the measurements are far from each other, then you have accuracy without precision. A good analogy for understanding accuracy and precision is to imagine a basketball player shooting baskets. If the player shoots with accuracy, his aim will always take the ball close to or into the basket. If the player shoots with precision, his aim will always take the ball to the same location which may or may not be close to the basket. A good player will be both accurate and precise by shooting the ball the same way each time making it in the basket.

L. Ahmad, S. S. Mahdi, *Satellite Farming*,
https://doi.org/10.1007/978-3-030-03448-1_1

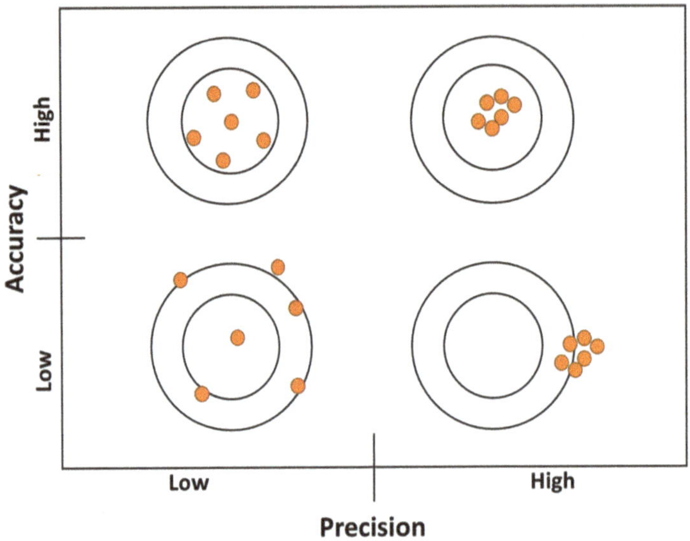

Fig. 1.1 Accuracy and precision

Accuracy and precision are used in context of measurement. Accuracy refers to the degree of conformity and correctness of something when compared to a true or absolute value, while precision refers to a state of strict exactness – how consistently something is strictly exact.

In other words, the precision of an experiment, object, or value is a measure of the reliability and consistency. The accuracy of an experiment, object, or value is a measurement of how closely results agree with the true or accepted value (Fig. 1.1). Both accuracy and precision are terms used in the fields of science, engineering, and statistics.

1.2 Comparison Chart

Accuracy versus precision comparison chart

	Accuracy	Precision
Definition	The degree of conformity and correctness of something when compared to a true or absolute value	A state of strict exactness – how often something is strictly exact
Measurements	Single factor or measurement	Multiple measurements or factors are needed
Relationship	Something can be accurate on occasion as a fluke. For something to be consistently and reliably accurate, it must also be precise	Results can be precise without being accurate. Alternatively results can be precise and accurate
Uses	Physics, chemistry, engineering, statistics, and so on	Physics, chemistry engineering statistics, and so on

Example

One can say that a measurement is accurate but not precise, precise but not accurate, or neither or both. An example of bad precision with good accuracy might be a lab refrigerator that holds a constant temperature of 38 °F. A temperature sensor is tested 10 times in the refrigerator. The temperatures from the test yield the temperatures of 37.8, 38.3, 38.1, 38.0, 37.6, 38.2, 38.0, 38.0, 37.4, and 38.3. This distribution shows no impressive tendency toward a particular value (lack of precision), but each value does come close to the actual temperature (high accuracy).

1.3 Target Comparison

Accuracy is the degree of correctness, while precision is how strict that correctness is (or isn't) – how reproducible results are. For this example, consider results from a round of target practice.

Arrows are fired at a target, and measurements are taken in relation to the bull's-eye at the center of the target. Accuracy describes how close the arrows are to the bull's-eye. The closer an arrow is to the bull's-eye, the more accurate the shot.

How precise the shots are depends on how often the arrows land near each other on the target. When all or most arrows are grouped tightly together, the shots fired can be considered precise since they all landed near the same spot, if not necessarily near the bull's-eye (Fig. 1.2). This is how results can indicate precision but not necessarily accuracy. However, it is important to note that it is not possible to reliably achieve accuracy without precision.

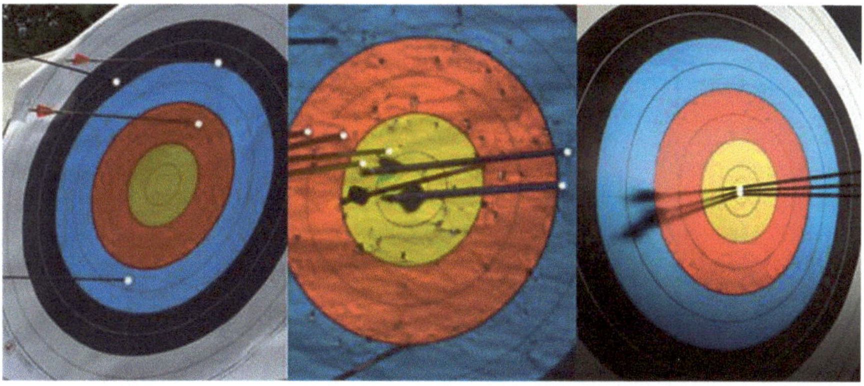

Fig. 1.2 The target at left displays neither accuracy nor precision, while the target in the middle displays some precision but little accuracy. Finally, the third target, at right, shows both accuracy and precision

1.4 Number of Measurements

Accuracy can be improved by taking repeat measurements and taking an average. (This assumes that errors are randomly above and below the true value to the same degree). Therefore an experiment with a low degree of precision can provide accurate values where appropriate statics are applied.

Conversely, precision cannot be improved by taking repeated measurements, but it is impossible to quantify precision without experimental repeats. The danger when evaluating an experiment is that some errors are not random. In this case, an experiment could yield inaccurate results yet be highly precise.

1.5 Quality

While a precise measurement may speak highly of an instrument's quality, an accurate reading will not reflect on the quality. Accuracy is an agreement of a measured value with an expected value. For example, a stopped clock will be accurate twice in day, but it will not be precise, i.e., able to consistently and accurately keep time throughout the day. In the case of a clock, how precisely it measures time matters a great deal and determines quality.

How to Remember?

- a**C**curate is **c**orrect (a bulls eye).
- p**R**ecise is **r**epeating (hitting the same spot but maybe not the correct spot)-, ç.

1.6 Bias

When we measure something several times and all values are close, they may all be wrong if there is a "bias." Bias is a systematic (built-in) error which makes all measurements wrong by a certain amount.

Examples of Bias

- The scales read "1 kg" when there is nothing on them.
- You always measure your height wearing shoes with thick soles.
- A stopwatch that takes half a second to stop when clicked.

In each case all measurements are wrong by the same amount. That is bias.

1.7 Degree of Accuracy

Accuracy depends on the instrument we are measuring with. But as a general rule, the degree of accuracy is **half a unit** each side of the unit of measure.

Example:

When an instrument measures in "1"s

any value between 6½ and 7½ is measured as "7"

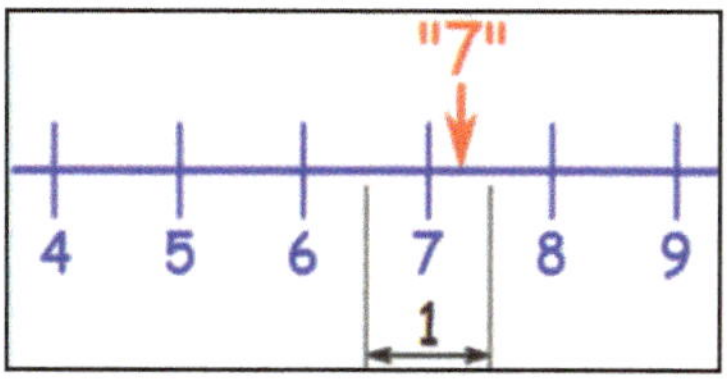

When an instrument measures in "2"s

any value between 7 and 9 is measured as "8"

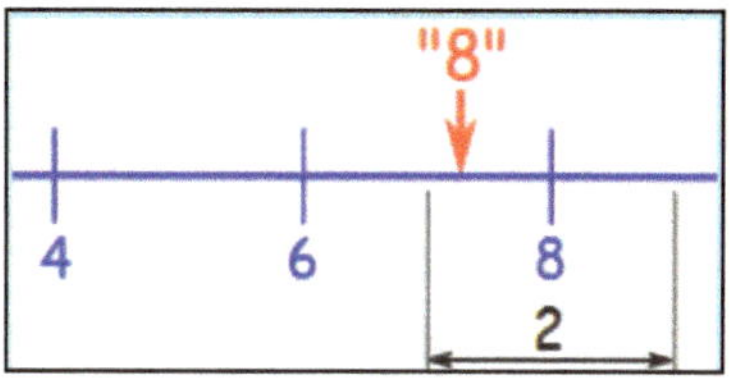

1.8 A Brief History of Precision Agriculture

Precision agriculture (PA) is no longer a new term in global agriculture. Since the first substantial PA workshop was held in Minneapolis in 1992, it has become the subject of numerous conferences worldwide. The impetus for the current concept of precision agriculture in cropping systems emerged in the late 1980s with the matching of grid-based sampling of soil chemical properties with newly developed variable-rate application (VRA) equipment for fertilizers. Using a compass and dead-reckoning principles, fertilizers were applied at rates designed to complement changes in soil fertility maps that had been created. Crop yield monitoring technologies were still in the research phase at this stage.

Around 1990, the NAVSTAR global positioning system (GPS) became available in a limited capacity for civilian use, and the opportunity for rapid and "accurate" vehicle location and navigation sparked a flurry of activity. Electronic controllers for VRA were built to handle this new positioning information, and crop yield monitors began to hit the commercial market. By 1993 the GPS was fully operational, and a number of crop yield monitoring systems were allowing the fine-scale monitoring and mapping of yield variation within fields. The linking of yield variability data at this scale with maps of soil nutrient changes across a field marked the true beginning of PA in broadacre cropping.

As yield monitoring systems were improved, it became evident that methods other than grid sampling for collaborative information would need to be developed. In many instances, grid sampling at the intensity required to correctly characterize variability in soil and crop parameters proved cost prohibitive, and, by the late 1990s, a "zonal" management approach had become a real option for management. This approach subdivides existing fields into zones of similar crop response and helps account for current limitations in data resolution while trying to maximize the benefits of PA for crop management.

New systems for measuring or inferring soil and crop parameters on a more continuous basis continue to be developed using both proximal (i.e., on ground-based platforms) and remote (i.e., aerial and satellite) platforms. Examples of these are soil EC, a measuring instrument, crop reflectance imaging, and crop quality sensors.

The success and potential for further success, observed in the grains industry, prompted other farming industries, particularly viticulture and horticultural crops, to adopt precision agriculture. Since the late 1990s, more and more research has been carried out in non-grain crops. Also, more emphasis is being placed on the environmental auditing capabilities of PA technology and the potential for product traceability. Advances in Global Navigation Satellite System (GNSS) technology since 1999 have also opened the door for machinery guidance, auto-steering, and controlled-traffic farming (CTF). CTF has provided sustainability benefits (such as minimization of soil compaction), economic benefits (by minimizing input overlap and improving timeliness of operations), and social benefits (such as reducing driver fatigue). As a result this form of PA technology has been showing swift adoption rates in the first decade of the twenty-first century.

1.9 Defining Precision Agriculture

Precision agriculture (PA) is also known as precision farming. Perhaps the easiest way to understand PA is to think of it as everything that makes the practice of farming more accurate and controlled when it comes to the growing of crops and raising livestock. A key component of this farm management approach is the use of information technology and a wide array of items such as GPS guidance, control systems, sensors, robotics, drones, autonomous vehicles, variable rate technology, GPS-based soil sampling, automated hardware, telematics, and software.

Precision agriculture or satellite farming or site-specific crop management is a farming based on observing, measuring, and responding to inter- and intra-field variability in crops.

Or

PA is an information- and technology-based farm management system to identity, analyze, and manage variability within fields for optimum profitability, sustainability, and protection of land resources.

Or

Precision application of technologies and input based on soil, crop weather, and market demands to maximize sustainable productivity and profitability.

Or

Precision forming is generally defined as an information and technology best for management system to identify, analyze, and manage variability within fields for optimum profitability, sustainability, and production of land resource.

In India, the average landholdings are very small even with large and progressive farmers. It is thus necessary to design sensed definition of precision farming in the context of Indian scenario.

Precision application of agricultural inputs based on soil weather and crop requirement to maximize sustainability and productivity and quality and profitability.

Aim of PA

Precision agriculture aims to optimize field level management with regard to:

1. Crop science by matching crop practices more closely to crop needs (e.g., fertilizer inputs)
2. Environmental protection: by reducing environmental risks and footprints of farming (e.g., limiting leaching of nitrogen)
3. Economics: by boosting competitive through more efficient practices (improved management of fertilizer usage and other inputs)

Objectives:

1. Site-specific application of inputs required to grow crops = improve crop productivity
2. Increase in the efficiency of agricultural resources utilization by crops = lower environmental impact
3. Automate and log farm operation data analysis efficiency and convenience

Many definitions of PA exist, and many people have different ideas of what PA should encompass. Here two definitions have been selected to illustrate the concept of PA in general but also specifically its application to broadacre cropping industries. The first definition comes from the US House of Representatives (US House of Representatives 1997).

Precision Agriculture May Be Defined as:

> "an integrated information- and production-based farming system that is designed to increase long term, site-specific and whole farm production efficiency, productivity and profitability while minimizing unintended impacts on wildlife and the environment."

The key to this definition is that it identifies PA as a "whole-farm" management strategy (not just for individual fields) that utilizes information technology and that the aim of management is to improve production and minimize environmental impact. It also refers to the farming system which in modern agriculture may include the supply chain from the farm gate to the consumer. This definition also distinguishes between agriculture and agronomy. While the PA philosophy has been expounded primarily in cropping industries, it is important to remember that precision agriculture can relate to any agricultural production system. These may involve animal industries, fisheries, and forestry, and in many cases, PA techniques are being implemented without being identified as such. For example, the tailoring of feed requirements to individual milkers depending on the stage of their lactation in a dairy enterprise.

The second definition narrows the PA philosophy of timely management of variation down to its implementation in cropping systems.

Site-Specific Crop Management (SSCM)

> "A form of PA whereby decisions on resource application and agronomic practices are improved to better match soil and crop requirements as they vary in the field."

This definition encompasses the idea that PA is an evolving management strategy (Fig. 1.3). The focus here is on decision-making with regard to resource use and not necessarily the adoption of information technology on farm (although many new technologies will aid improved decision-making). The decisions can be in regard to changes across a field at a certain time in the season or changes through a season or seasons. The inference is that better decision-making will provide a wide range of benefits (economic, environmental, and social) that may or may not be known or measurable at present.

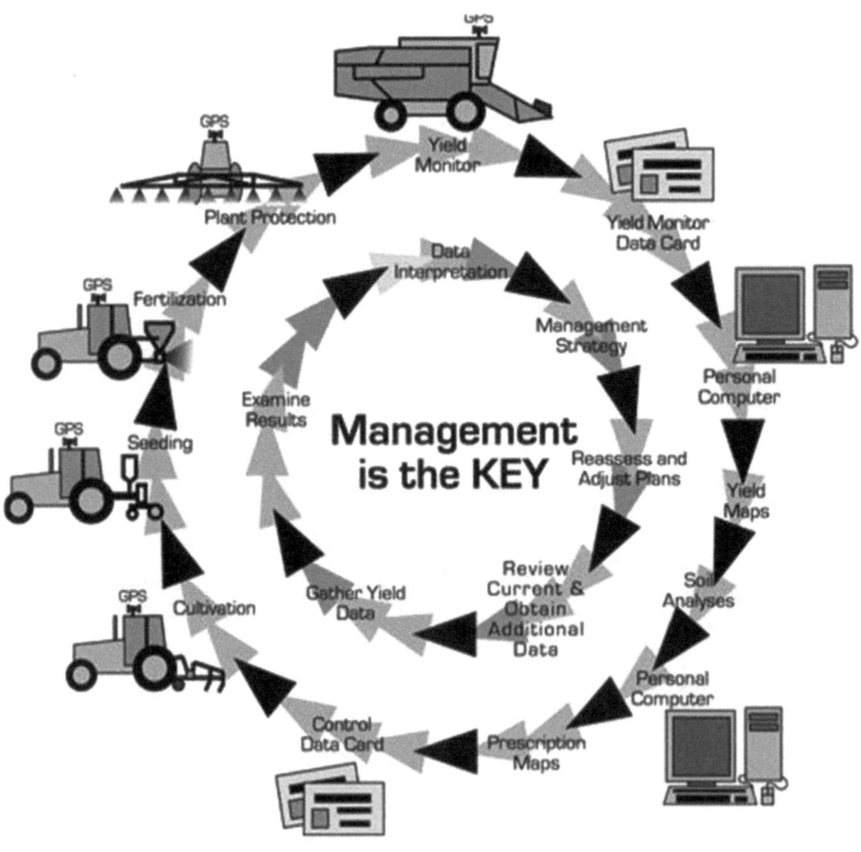

Fig. 1.3 Precision agriculture: a comprehensive approach (Banu 2015)

To further expand the concept, SSCM can be considered as the application of information technologies, together with production experience, to:

1. Optimize production efficiency.
2. Optimize quality.
3. Minimize environmental impact.
4. Minimize risk.

– All at the site-specific level

This is not a particularly new concept in agriculture with essays on this topic dating from the early eighteenth century. What is new is the scale at which we are able to implement these aims. Prior to the industrial revolution, agriculture was generally conducted on small fields with farmers often having a detailed knowledge of their production system without actually quantifying the variability. The movement toward mechanical agriculture, and the profit margin squeeze, has resulted in the latter half of the twentieth century being dominated by large-scale uniform "average" agricultural practices. The advance of technology in the late twentieth and early twenty-first centuries has allowed agriculture to move back toward site-specific agriculture while retaining the economies of scale associated with "large" operations.

Some Misconceptions
Like many new concepts, PA carries with it some misconceptions (Taylor and Whelan 2005).

- PA is often confused with yield mapping. Yield mapping is a tool that is one of the first steps toward implementing a SSCM strategy.
- PA is sometimes misinterpreted as sustainable agriculture. PA is a tool to help make agriculture more sustainable; however, it is not the total answer. PA aims at maximum production efficiency with minimum environmental impact. Initially it was the potential for improved productivity (and profitability) that drove the development of SSCM as a form of PA. In recent years the potential for this technology as a tool for environmental auditing of production systems has become more obvious. However environmental auditing is not environmental management. The large amount of fine-scale data being collected in a SSCM system can be used for on-farm environmental risk assessment and incorporated into a whole-farm plan to help viability in the long term.
- Finally, machinery guidance and auto-steer systems are examples of the successful adoption of new technology on farms. However, these again are tools that help with SSCM. By themselves they are not PA.

1.10 Variability and the Production System

SSCM is dependent on the existence of variability and broadly speaking "variability in production = SSCM opportunity." Having said this, the type, magnitude, and distribution pattern of variability are also important. There are generally two types of variability to be considered, spatial or temporal. Spatial variability occurs over a measurable distance; temporal variability occurs over a measurable time period. The difference between the low and high values of a measured property defines the magnitude in both types of variability. The distribution pattern maps how variability is changing in either the space or time dimension.

The management implications of these aspects of variability are diverse and fundamentally linked to the production property being measured. However there are a few simple generalizations that are worth keeping in mind. The observed magnitude in the variability should be related to a benchmark level below which it would be uneconomical to attempt to manage. It is important to note that the costs used to calculate these benchmarks are presently considered from a short-term economic perspective. If we were able to express environmental benefits in a fiscal sense, then in some instances, areas with a small magnitude of variation in production may be viable for SSCM management.

The distribution pattern of the variability needs to be considered relative to the options for management intervention. In spatial terms, the pattern should be considered in relation to the smallest unit of treatment applicable (e.g., the size and reaction time of VRA fertilizer application gear). In temporal terms, the pattern should be considered in terms of the impact on important management stages of the growing season (or the whole season if relevant).

If spatial variability does not exist, then a uniform management system is both the cheapest and most effective management strategy. In cropping situations the magnitude of temporal variability may appear much greater than spatial variability. If the impact of temporal variability on production overwhelms the impact of spatial variability, then careful consideration needs to be given to whether a uniform or differential management strategy is the optimal risk aversion strategy.

Based on these considerations, SSCM is at present operating on a zonal rather than a completely site-specific basi*s* (Fig. 1.4). As our ability to measure variability improves, the capital cost of VRA technology decreases, and the environmental value is factored in; SSCM will begin to approach a truly site-specific management regime.

1.11 Need for Precision Agriculture

The potential of precision farming for economic and environmental benefits could be visualized through reduced use of water, fertilizers, herbicides, and pesticides besides the farm equipment. Instead of managing an entire field based upon some hypothetical average condition, which may not exist anywhere in the field, a

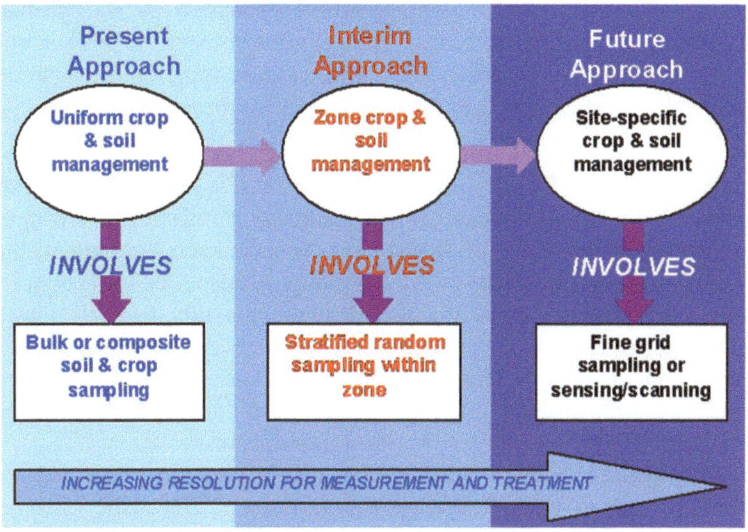

Fig. 1.4 The evolving timeline of SSCM from a uniform to a totally site-specific approach (GRDC 2009)

precision farming approach recognizes site-specific differences within fields and adjusts management actions accordingly (Singh et al. 2008). Farmers usually are aware that their fields have variable yields across the landscape. These variations can be traced to management practices, soil properties, and/or environmental characteristics. A farmer's mental information database about how to treat different areas in a field required years of observation and implementation through trial and error.

Today, that level of knowledge of field conditions is difficult to maintain because of the larger farm sizes and changes in areas cultivated due to annual shifts in leasing arrangements. Precision agriculture offers the potential to automate and simplify the collection and analysis of information. It allows management decisions to be made quickly.

1.12 Objectives of SSCM

At the beginning of this introduction, SSCM was defined in terms of four main objectives. The success of a SSCM strategy will depend on how each or all of these objectives are met.

Optimizing Production Efficiency

In general the aim of SSCM is to optimize returns across a field. Unless a field has a uniform yield potential (and therefore a uniform yield goal), the identification of variability in yield potential may offer possibilities to optimize production quantity at each site or within each "zone" using differential management. The initial emphasis should be on optimizing the agronomic response to the manageable input with the most impact on production and costs. In the absence of any clear environmental benefits, this will be achieved by differentially applying inputs so that the marginal return = marginal cost at each site or zone in the paddock.

Optimizing Quality
In general, production efficiency is measured in terms of a yield (quantity) response, mainly because yield and biomass sensors are the most reliable and commonplace sensors. In the past few years, the first attempts to commercialize grain quality sensors have been made, and on-the-go grain protein/oil sensors are now commercially available. The ability to site-specifically collect grain quality data will allow growers to consider production efficiency from the perspective of either yield, quality, or a yield x quality interaction. Many inputs will impact on quality as well as quantity. In production systems where quality premiums exist, this may alter the amount of input required to optimize profitability and agronomic response.

In some product markets, where strong quality premiums/penalties are applied, a uniform approach to quality properties may be optimal. The quality of some agricultural commodities is greatly increased by decreasing the variability in production, e.g., wine grapes or malting barley. If quality premiums more than offset yield loss, then growers may prefer to vary inputs to achieve uniform production quality (and minimize variability) rather than optimize productivity.

Minimizing Environmental Impact
If better management decisions are being made to tailor inputs to meet production needs, then by default there must be a decrease in the net loss of any applied input to the environment. This is not to say that there is no actual or potential environmental damage associated with the production system; however, the risk of environmental damage is reduced.

SSCM, coupled with VRA technology, provides producers with a means to not only quantify the amount and location of any input application but also to record and map applications. This gives producers physical evidence to contest any claims against negligent management or alternatively provide information on "considerate" practices to gain market advantage. A by-product of improved information collection and flow is a general improvement in the producer understanding the production system and the potential implications of different management options.

Minimizing Risk
Risk management is a common practice today for most farmers and can be considered from two points of view – income and environmental. In a production system, farmers often practice risk management by erring on the side of extra inputs, while the unit cost of a particular input is deemed "low." Thus a farmer may put an extra spray on, add extra fertilizer, buy more machinery, or hire extra labor to ensure that the product is produced/harvested/sold on time, thereby guaranteeing a return. Generally minimizing income risk is seen as more important than minimizing environmental risk, but SSCM attempts to offer a solution that may allow both positions to be considered in risk management. This improved management strategy will

come about through a better understanding of the environment–crop interaction and a more detailed use of emerging and existing information technologies (e.g., short- and long-term weather predictions and agro-economic modeling).

The more that is known about a production system, the faster a producer can adapt to changes in his own production and in external market forces. For example, accurate mid-season yield predictions may give a grower more room to move with forward selling options.

1.13 Practical Implementation of SSCM

The SSCM cycle is illustrated in Fig. 1.5. Each node in the cycle will form the theme for subsequent chapters in this series; however, a short introduction is given here. It is important to remember that SSCM is a continuous management strategy. Initially some form of monitoring and data analysis is needed to form a decision. However it is just as important to continue to monitor and analyze the effect of the decision and feed this information into subsequent management decisions.

Geo-referencing
The truly enabling technology of SSCM is in its present form. Global Navigation Satellite Systems (GNSS) (of which the GPS is the most widely used at present) are now commonplace on many farms. Receivers range in accuracy from 10–20 m to 2–3 cm, in price from $200 to $60,000, and in application from crop monitoring and yield mapping to auto-steer systems. The technology continues to improve and the price of receivers to decrease.

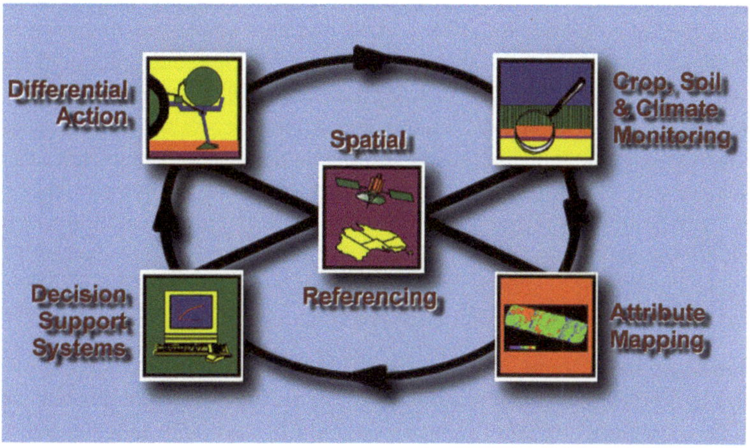

Fig. 1.5 The SSCM cycle indicating spatial referencing as the enabling technology that drives the other parts of the cycle

The ability to geo-reference activities gives producers the option to map and visually display farm operations. This provides insights into both production variability and inefficiencies in crop production and farm operations. In the past few years, more advanced systems have become more common on farms as growers embrace guidance and auto-steer technologies. These permit machinery to drive along repeatable tracks as well as reduce driver fatigue and permit more timeliness of operations.

Crop, Soil, and Climate Monitoring
Many sensors and monitors already exist for in situ and on-the-go measurement for a variety of crops, soil, and climatic variables. These include yield sensors, biomass and crop response sensors (aerial- and spaceborne multi- and hyper-spectral cameras), radio or mobile phone networked weather stations, soil apparent electrical conductivity (ECa) sensors, and gamma-radiometric soil sensors to name a few.

The other challenge for SSCM is to adapt in situ sensors and develop new on-the-go sensors. While the commercial potential of these sensors will mean that private industry will be keen to take up the engineering aspects of research and development, research bodies have an important role to play in the development of the science behind the sensors. Market concerns often lead private industry to sell sensors prematurely to ensure market share. This may lead to substandard sensors and a failure to adequately realize the potential of the sensor. Agricultural scientists also need to continue to assess which and how multiple crop and production indicators can be measured.

Attribute Mapping
Crop, soil, and climate sensors often produce large, intensive data sets. The observations are usually irregularly spaced and need to be "cleaned" and interpolated onto a surface to permit analysis. For several decades geostatisticians have been researching ways of describing and representing spatial data that accurately represents the raw data. Historically most of this has been done with sparse data sets. The data sets being generated by SSCM technologies have produced new challenges for mapping, but most of these are now well understood within the PA community although answers are generally poorly disseminated through the wider agricultural community.

Software for mapping and displaying data from different sources on a common platform is improving annually. The development of geographical information systems (GIS) specifically for agriculture is allowing this to occur; however, the adaptation and adoption of this technology for use in SSCM on individual farms are still in its infancy. The main issues still to be resolved are the development of a user-friendly advanced data filtering system and the determination of initial and future sampling schemes to ensure that the variability of the system is properly characterized.

Decision Support Systems
Techniques for data presentation and storage, such as geographical information systems (GIS), developed in other industries should be relatively easily applied, with some modification, to agriculture. However decision support systems (DSS) are not so flexible, and it is in this area that much research needs to be done. Decision support systems use agronomic and environmental data, combined with information on

possible management techniques, to determine the optimum management strategy for production. Most commercial DSS are based on "average" crop response across a field. The majority of engineering companies currently supplying SSCM technology are currently not producing DSS to support the differential use of their equipment in a production system. Therefore the onus is falling on individual industry bodies, and to lesser extent government agencies, to fill the gap. Initially it may be sufficient to adapt existing agricultural DSS such as WHEATMAN, COTTONLOGIC, or APSIM to site-specific situations. In the long run, however, a DSS that is able to site-specifically model plant-environment interactions in terms of yield and quality will be needed. This will need to be flexible enough to incorporate a variety of sensor-gathered data, accept feedback from other parts of the SSCM cycle, and be able to conform to standards such as ISO 9000/14000.

Differential Action

The differential application of inputs using VRA technology is essentially an engineering problem. Due to the commercial potential of VRA technology, much of this engineering development is being driven by the private sector. The main input required for VRA implements is accurate information on required application rates and associated locations or times for the applications. VRA equipment should also record the actual application procedure for quality control. The differential application technology was probably the best developed part of the SSCM cycle in the late 1990s, and development of new methods for differential application remains a project of many research and commercial entities around the globe.

Like GPS receivers, VRA equipment is becoming more user friendly, more cost effective, and more common especially in broadacre agriculture. The biggest barrier to adoption is the lack of information from a DSS on where, and by how much, inputs should be varied.

1.14 Steps in Precision Farming

The basic steps in precision farming are (Ali 2015):

(a) Assessing variation
(b) Managing variation and
(c) Evaluation

The available technologies enable us in understanding the variability, and by giving site-specific agronomic recommendations, we can manage the variability that makes precision agriculture viable. And finally evaluation must be an integral part of any precision farming system.

(a) **Assessing Variability**

Assessing variability is the critical first step in precision farming. It is clear that one cannot manage what one does not know, factors and the processes that regulate or control the crop performance in terms of yield vary in space and time. Quantifying the variability of these factors and processes and determining when and where different combinations are responsible for the spatial and temporal variation in crop yield is the challenge for precision agriculture.

Techniques for assessing spatial variability are readily available and have been applied extensively in precision agriculture. The major part of precision agriculture lies in assessing to spatial variability. Techniques for assessing temporal variability also exist, but the simultaneous reporting of a spatial and temporal variation is rare. We need both the spatial and temporal statistics. We can observe the variability in yield of a crop in space, but we cannot predict the reasons for the variability. It needs the observations at crop growth and development over the growing season, which is nothing but the temporal variation. Hence, we need both the space and time statistics to apply the precision farming techniques. But this is not common to all the variability/factor that dictate crop yield. Some variables are more produced in space rather than with time, making them more conducive to current forms of precision management.

Steps in Accessing Variability:

1. Surveys
2. Interpolation of point samples
3. High-resolution sensing
4. Modeling

(b) **Managing Variability**

Once variation is adequately assessed, farmers must match agronomic inputs to known conditions employing management recommendations. Those are site-specific and use accurate applications control equipment.

We can use the technology most effectively. In site-specific variability management, we can use GPS instrument, so that the site specificity is pronounced and management will be easy and economical. While taking the soil/plant samples, we have to note the sample site coordinates, and further we can use the same for management. This results in effective use of inputs and avoids any wastage, and this is what we are looking for.

The potential for improved precision in soil fertility management combined with increased precision in application control make precise soil fertility management as attractive but largely unproven alternative to uniform field management. For successful implementation, the concept of precision soil fertility management requires that within-field variability exists and is accurately identified and reliably interpreted that variability influences crop yield, crop quality, and for the environment. Therefore inputs can be applied accurately.

The higher the spatial dependence of a manageable soil property, the higher the potential for precision management and the greater its potential value. The degree of difficulty, however, increases as the temporal component of spatial variability increases. Applying this hypothesis to soil fertility would support that phosphorus and potassium fertility are very conducive to precision management because temporal variability is low. For N, the temporal component of variability can be larger than its spatial component, making precision N management much more difficult in some cases.

Steps in Managing the Variability:
1. Precision soil fertility management
2. Precision pest management
3. Crop management
4. Water management
5. Soil management

(c) **Evaluation**

There are three important issues regarding precision agriculture evaluation.

- Economics
- Environment and
- Technology transfer

The most important fact regarding the analysis of profitability of precision agriculture is that the value comes from the application of the data and not from the use of the technology. Potential improvements in environmental quality are often cited as a reason for using precision agriculture. Reduced agrochemical use, higher nutrient use efficiencies, increased efficiency of managed inputs, and increased production of soils from degradation are frequently cited as potential benefits to the environment. Enabling technologies can make precision agriculture feasible; agronomic principles and decision rules can make it applicable, and enhanced production efficiency or other forms of value can make it profitable.

The term technology transfer could imply that precision agriculture occurs when individuals or firms simply acquire and use the enabling technologies. While precision agriculture does involve the application of enabling technologies and agronomic principles to manage spatial and temporal variability, the key term is manage. Much of the attention in what is called technology transfer has focused on how to communicate with the farmer. These issues associated with the managerial capability of the operator, the spatial distribution of infrastructure, and the compatibility of technology to individual farms will change radically as precision agriculture continues to develop (Pierce and Nowak 1999).

1.15 Conclusions

Precision agriculture is a management philosophy, encompassing the use of advances in information technology in agriculture. In 10–15 years' time, it is likely (and hoped) that SSCM as a form of PA, with its associated technologies and methodologies, will be simply considered as standard cropping practice. But no matter how technology and methodologies change and adapt over time, SSCM will still be driven by the central philosophies of improved production efficiency, reduced environmental impact, and risk minimization.

References

Ali, J. (2015). *Role of precision farming in sustainable development of hill agriculture.*

Banu, S. (2015). Precision agriculture: Tomorrow's technology for today's farmer. *Journal of Food Processing & Technology, 6*, 468. https://doi.org/10.4172/2157-7110.1000468.

GRDC. (2009). Precision agriculture fact sheet (pp 1–9).

Pierce, F. J., & Nowak, P. (1999). Aspects of precision agriculture. *Advances in Agronomy, 67*, 1–85.

Singh, A. K., Tripathy, R., & Chopra, U. K. (2008). Evaluation of CERES-Wheat and CropSyst models for water–nitrogen interactions in wheat crop. *Agricultural Water Management, 95*(7), 776–786.

Taylor, J., & Whelan, B. (2005). *A general introduction to precision agriculture*. Australian Center for Precision Agriculture.

United States Geological Survey. (n.d.). Retrieved from: https://www2.usgs.gov/faq/node/2858. U.S. House of representatives. (1997).

Chapter 2
Components of Precision Agriculture

Abstract Precision agriculture (PA) or precision farming (PF) has witnessed unprecedented growth in the last decade, especially in countries such as the United States, Germany, Canada, and others. While the rest of the world has been relatively slow in embracing precision agricultural practices, the change is coming. The chapter discusses different components of precision agriculture.

Keywords Components · Data management · Soil sampling · Yield monitoring · Variable rate application · Variable rate technology

2.1 Components of Precision Agriculture

Ever since the man appeared on the earth, he has been harnessing the natural resources to meet his basic requirements. The term "precision farming" describes the integration of GIS and GPS tools to provide an extensive amount of detailed information on crop growth, crop health, crop yield, water absorption, nutrient levels, topography, and soil variability (Adrian et al. 2005). It precisely establishes various operations, such as the best tillage, application of fertilizer, sowing, irrigation, harvesting, etc., and turns traditional extensive production to intensive production according to space variable data. It not only utilize fully resources, reduce investment, decrease pollution of the environment, and get the most of social and economic efficiency but also makes farm products, the same as industry, become controllable and be produced in standards and batches (Khan 2013). Thus, specific objectives of precision agriculture are to increase profitability, increase production, reduce variable costs, reduce erosion, reduce the environmental impact of chemicals, track and monitor the use of chemicals, and manage large farms (Atherton et al. 1999; Olson 1998). However, precision farming has been confined to developed countries. Land tenure system, smaller farm size (<1 ha), and crop diversity have limited the scope of precision farming in India. However, there is a wide scope for precision farming in irrigated/commercial/fruit and vegetable crops/high value crops. It is apparent that there is a tremendous scope for precision farming in India

L. Ahmad, S. S. Mahdi, *Satellite Farming*,
https://doi.org/10.1007/978-3-030-03448-1_2

as well, and it is necessary to develop database of agriculture resources, which will act as decision support system at the farm (<1 ha) level.

Generally, three major components of precision agriculture are information, technology, and management. Base on these three principles, we can define PA in different ways. Precision farming is information-intense. A lot of data are required to generate treatment maps, and many techniques are underdevelopment or have been developed during the last decade. Earlier, precision agriculture encompassed three "Rs": the right time, the right amount, and the right place (Robert et al. 1994). Later, the International Plant Nutrition Institute added another "R" to that list, "the Right Source," and more recently an additional "R," the right manner, has also been proposed (Khosla 2008).

Precision agriculture is a management strategy that uses information technologies to bring data from multiple sources to bear on decisions associated with crop production (National Research Council 1997). It allows for the management of spatial and temporal variability within a field, reduction of costs, improvement of yield quantity and quality, and reduction of environmental impacts.

In other hand, PA can be defined as a set of technologies that have helped propel agriculture into the computerized information-based world and is designed to help farmers get greater control over the management of farm operations (Gandonou 2005). Precision farming is based on several separate technologies, yet interdependent parts, which together form the basis for individual management systems.

Precision farming technology (PFT) is designed to provide information and data to assist farmers when making site-specific management (SSM) decisions. Demand for information about technologies to manage agricultural production systems increased with the advent of yield monitors and global positioning systems, etc., and improvements in computing power and data management (Swinton and Lowenberg-DeBoer 1998). The following section outlines the most prevalent systems that are used in PA.

The key elements required for generation, dissemination, and adoption of precision farming techniques, therefore, include information, technology, and managements.

Information is needed about crop characteristics, soil properties (topography, fertility status, texture, moisture content/retention, tillage needs, salinity, waterlogging, etc.), incidence of pests (insects, diseases, weeds, and others), weather/climatic conditions, other biotic and abiotic stresses, plant growth response, harvest and postharvest handling, marketing and market intelligence, socioeconomic conditions of farmers, etc. Detailed information, so collected, can be used to create different kinds of maps of the farms/villages/regions (e.g., different soil characteristics, groundwater, pest incidence, weed distribution, topography, environmental pollution, etc.) which can help the farmers in using the available information at each step while making site-specific decisions for performing different farm operations. Planners can use these maps for regulating the supplies of inputs in different areas.

Technologies are rapidly evolving, and farmers must keep up with the changes that can help in increasing productivity and profitability. Use of remote sensing and geographic information system (GIS), auto-analysers, sensors, computers along with appropriate software, etc. can help in precisely identifying areas of nutrient deficiencies and other biotic and abiotic stresses, etc., and quantifying the quantification of the economic significance of the soil-water-fertilizer-pest-crop-related constraints and their environmental impacts at the farm/village/region levels. They can provide useful guidance for adopting the systems of integrated management of soil health, nutrients, pests, water, energy, and different crop genetic resources. By using sensors postharvest quality of the produce can be monitored and enhanced.

Efficient **management** of the suite of technologies is of paramount importance. This will combine information obtained and the available technology into a comprehensive management system. Without proper management precision, crop production will not be possible. Farmers must know how to interpret available information, how to utilize the technology, and how to make sound production and decisions.

The complexity of agriculture and diverse expertise needed for the development and dissemination of knowledge-intensive precision farming technologies require multidisciplinary efforts of agronomists, plant breeders, soil scientists, agrometeorologists, entomologists, plant pathologists, weed scientists, biotechnologists, economists, extension workers, and farmers. Paradigm shifts are needed in the systems of generation, transfer, and adoption. Systems approach needs to be followed to integrate, synthesize, and process information, and develop computer-simulated crop production models. User-friendly computer software can in turn guide the farmers in making appropriate decisions for performing site-specific farm operations. User-oriented cost-effective technologies generated through such efforts will help in ensuring the adoption of precision farming technologies by the farmers.

2.2 Spatial Data Management

Field is not one uniform unit. In principle, all parameters and farming practices can be a part of a site-specific resource management.

2.2.1 Geographical Positioning

Precision farming, as we know it today, started in the late 1980s with the advent of the global positioning system (GPS) into the agricultural sector, when it was realized how much farming data were spatially related (Blackmore 2003). Geographical positioning with GPS or DGPS (differential GPS) is the "backbone" of most PF practices. It is essential for most site-specific practices that a specific action is recorded and positioned in order to use the information for future treatments. With a positioning system, it is possible to gather information site-specifically. The global positioning system (GPS) makes possible to record the in-field variability as geographically encoded data (Nemenyi et al. 2003). An alternative to the GPS system, the Russian GLONASS satellite system is developed, which in principle is similar to the GPS system and based on several satellites.

The global positioning system (GPS) is one of the key technologies that make precision agriculture possible. GPS receivers with sufficient accuracy for yield mapping, grid sampling, variable rate application, and other precision activities are available at moderate cost. GPS receivers, which provide accurate, geo-referenced position information, are often used with combine yield monitors, scouting equipment, or variable rate application machinery (Stombaugh et al. 2002). Several factors, including the satellite and receiver clock limitations, ephemeris variation, satellite configuration, atmospheric interference, multipath, and selective availability (SA) can cause errors in GPS position information.

The most common way to counteract GPS errors is by using differential GPS or DGPS. The differential global positioning system (DGPS) is an integration of space-based and ground-based segments that together comprise a radio-navigation facility (Shearer et al. 1999). In a DGPS system, a GPS receiver is placed at an accurately known location. This base station receiver will calculate GPS errors by comparing its actual location to the location computed from the GPS signals. This error information is sent to the rover receiver, which uses it to correct the position information it computes from the GPS signals. DGPS corrections can be broadcast by tower-based or satellite-based systems (Fig. 2.1).

By using the geographical positioning system, it is possible to locate the information for further analysis and make a visual presentation in a GIS system. Precision farming provides a new solution using a systems approach for today's agricultural issues such as the need to balance productivity with environmental concerns. It is based on advanced information technology. It includes describing and modeling variation in soils and plant species and integrating agricultural practices to meet site-specific requirements. It aims at increased economic returns, as well as at reducing the energy input and the environmental impact of agriculture.

2.2.2 *Geographical Information System*

In order to save your geographical position and site-specific field characteristics, it is an indispensable to have a geographical information system (GIS) platform to save and handle spatial information. The GIS created by computing background makes possible to generate complex view about our fields and to make valid agrotechnological decisions. Most GIS systems are reliable but also fairly expensive depending on the features and capabilities of the program.

A GIS is a set of computer tools that allows one to work with data that are tied to a particular location or spatially mapped area on the earth. A GIS is a database that is specifically designed to work with map data (Price 2006). GIS is the merger of attribute and a geographic database. This allows for multiple detailed data to be graphically depicted on a map and utilized for decision-making. Farmers have long since utilized maps for data collection and decision-making. However, the difference with applying the advanced technology of GIS is that the GIS map exhibits "intelligence" where you can ask a question and get an answer. GIS technology applica-

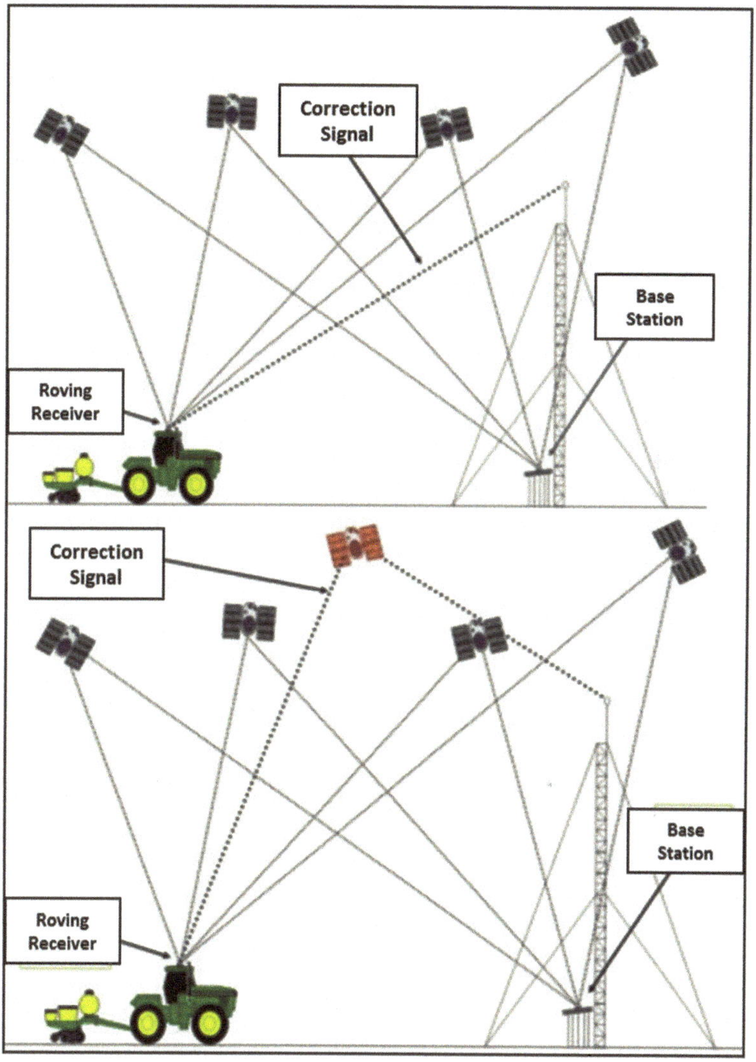

Fig. 2.1 DGPS tower-based (left) and satellite-based systems

tion is influencing decision-making because it avoids the shortcomings of traditional maps, allows for the rapid computer analysis, and applies sophisticated data structures. GIS is used as a data synthesizing and decision-support tool in many fields (Mickle 2009). GIS is a spatial decision support system, not a decision-making system.

A GIS software program enables the farmer to assemble and organize different sources of site-specific and geographical data information in various layers. Each layer in GIS is termed "coverage" and consisting of topologically linked geographic features and associated data.

2.2.3 *Remote Sensing*

Remote sensing is the science and art of acquiring information about the earth's surface without actually coming in contact with it. This is done by recording energy, which is either reflected or emitted from the earth's surface. The information recorded is then processed and analyzed, and the information is used to develop a prescription map that can be used in a variable rate application (Grisso et al. 2011). This technology can be used to obtain various layers of information about soil and crop conditions. It allows detection and/or characterization of an object, series of objects, or the landscape without having the sensor in physical contact with the soil. It uses aerial or satellite imaging to sense crop vegetation and identify crop stresses and injuries or pest infestation. So far satellite images have mainly been used to produce maps and predict yield potentials on large areas. With higher resolution and near-infrared and red sensing, it is possible to map the relative reflectance from the crop in a small scale.

2.3 Soil Sampling and Mapping

A variety of different soils exist across any given field. Also, a field may contain a low level of one nutrient and a high level of another nutrient. Soil sampling and testing are the best ways to identify these differences and to adjust liming and fertilization practices.

Both grid and directed sampling are used to describe soil properties for management of variable rates of fertility. Grid soil sampling involves sampling at points on a square grid throughout a field. An alternative to grid sampling is directed sampling, which involves dividing a field into regular or irregular management zones based on features such as soil type, topography, or past yield performance. In either system, collect at least five or six subsamples from each cell or zone and mix them into one sample container. The soil samples are collected from each section and then forwarded to a laboratory for analysis often guided by the advisory service. It is absolutely necessary to take care to assure that the soil sample you send to the laboratory accurately represents the area sampled (Thom et al. 1997). When soil fertility data is returned from the soil testing lab, you can enter the data into your mapping program to create fertility maps.

Gathering of soil data from soil samples to create a soil map is a relatively expensive and time-consuming task but have also valuable information for later treatments with fertilizers and design of decision support systems. A soil analysis can identify some of the limiting factors regarding yield and growth potential for the specific crop on a specific spot in the field. Soil test results should be included in a record system for each production field on a farm, along with the amounts of lime and fertilizer applied each year, the crops grown, and the yields obtained.

2.4 Yield Monitoring and Mapping

Yield monitors and maps form a very important part of a PF system. Yield monitors are a logical first step for those who want to begin practicing SSCM. The yield monitor which predicts yield from other variables is intended to give the user an accurate assessment of how yields vary within a field. With a yield monitor, a producer also can conduct on-farm variety trials or weed control evaluations without the need of a weigh wagon. Such on-farm comparisons help producers fine-tune crop production practices to their soils. Yield monitors are a combination of several components. They typically include several different sensors and other components, including a data storage device, user interface (display and key pad), and a task computer located in the combine cab, which controls the integration and interaction of these components. The sensors measure the mass or the volume of grain flow (grain flow sensors), separator speed, ground speed, grain moisture, and header height. The sensors are interfaced with both analog to digital and direct digital inputs. Yield is determined as a product of the various parameters being sensed (Shearer et al. 1999).

Grain volume flow sensor is prone to errors on slopes as the grain can slip to one side and block the beam. By adding slope sensors, this has been accounted for. Yield monitoring values can be exported to a personal computer and stored in nonvolatile memory for further analysis or printing via specialized software packages. Yield data were recorded and analyzed over 10 years, resulting in a rich data set for both spatial and temporal trend analysis 5. Variable-rate rechnology (VRT) is the equipment with the ability to change the application rate and mixture as it moves across the field. VRT adjust pesticide, herbicide, and seeding application rates to match soil potentials or problem areas (Adamchuck and Mulliken 2005). Roberts et al. (2001) reported that N losses to the environment were lower with VRT than with uniform rate application (URA), except on fields with little spatial variability and management information.

2.5 Differential Global Positioning System

The differential global positioning system (DGPS) is an integration of space- and ground-based segments that together comprise a radio-navigation facility. Initially developed for national security interests, a portion of the DGPS system is available to civilian users (Fig. 2.2). When yield data are used with information generated by a DGPS receiver, a producer can generate yield maps that provide a quick visualization of crop performance within a particular crop production unit. Ultimately, any increased profit realized from incorporating a yield monitor into an operation will come from changes in management practices that result from the identification of problem areas using such yield maps. With DGPS, the benefits of using a yield monitor are even more evident (Shearer et al. 1999).

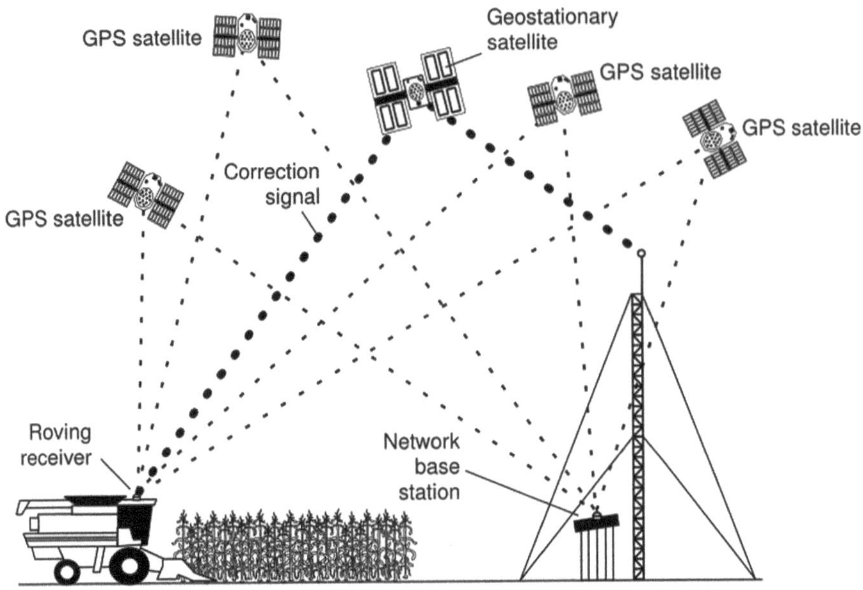

Fig. 2.2 Diagrammatic representation of DGPS

2.6 Variable Rate Application and Technology

The concept of PF has the potential of aiding farmers to use nutrients and chemicals in an efficient way according to crop needs at a given site on the field and thereby make farming more economically competitive and sustainable.

Application (VRA) of nutrients has the potential to improve the specific crop quality in a number of rotations, for example, to obtain certain protein content in cereals, in particular wheat for bakery flour and barley for malting. Variable-rate technology (VRT) is the equipment with the ability to change the application rate and mixture as it moves across the field. VRT adjust pesticide, herbicide, and seeding application rates to match soil potentials or problem areas (Adamchuck and Mulliken 2005). Six main sources of errors of yield measurement that have been identified are unknown crop width entering the header during harvest, time lag of grain through the threshing mechanism, the inherent wandering error from the GPS, surging grain through the combine transport system, grain losses from the combine, and sensor accuracy and calibration (Blackmore and Marshall 1996). Yield mapping is one application of data analysis where the process of observation, interpretation, evaluation, and implementation can be applied. Yield mapping software can be purchased as a part of an overall investment in a combine harvester or separately.

The two basic technologies for VRA are map-based and sensor-based. Map-based VRA adjusts the application rate based on an electronic map, also called a prescription map. Using the field position from a GPS receiver and a prescription map of desired rate, the concentration of input is changed as the applicator moves through the field. Canopy management can be conducted with a combination of crop sensors and real-time modeling. Sensor-based VRA requires no map or positioning system. Sensors on the applicator measure soil properties or crop characteristics “on the go.” Based on this continuous stream of information, a control system calculates the input needs of the soil or plants and transfers the information to a controller, which delivers the input to the location measured by the sensor (Grisso et al. 2011).

2.7 Section Control

Precision agriculture technology has evolved in such a manner that it provides farmers with new and innovative ways to possibly improve profitability. One of these ways is a new approach to the application of liquid chemicals and other inputs known as automatic section control. Automatic section control on a sprayer has the ability to selectively manage input application across the spray boom. This technology utilizes a global positioning system (GPS) to locate the position of the sprayer within the field and then records the areas covered. If the sprayer traverses an area previously covered, it can automatically turn the appropriate section off, therefore eliminating over application. In addition, automatic section control can manage chemical application in undesirable areas such as point rows, waterways, and during headland turns. The largest benefit associated with automatic section control is the reduction in overlapped areas especially prevalent on irregular-shaped fields. As a result, this new technology has the potential to increase profits due to reducing input costs. Environmental benefits are also possible due to the ability to manage buffer zones and protect sensitive areas in and around the field. Furthermore, improved efficiency can occur if coupled with a navigational aide such as light bar or autosteer (Shockley 2010). Once a prototype was created and farm trials were conducted, Dillon et al. (2007) examined the economic implications of utilizing automatic section control. Mooney et al. (2009) conducted the most recent economic analysis, and determined automatic section control became profitable at input saving levels of 11% or above.

2.8 Guidance Systems

There are two classifications of guidance systems for agricultural production: guidance aides and autonomous systems. Guidance aides are devices that provide guidance information to the operator and still require the operator to fully control the machines operations. The intent of the autonomous system is to free up the operator of the machine from the guidance task and thus improve operating efficiency.

2.8.1 Guidance Aid

The term guidance aide refers to devices that provide guidance information to the operator but do not attempt to replace the operator (Ima and Mann 2004). One of the most broadly adopted and more recently developed methods to reduce guidance error is the light bar technology. The main function of light bar is to assist the machine operator in driving, thus decreasing operator fatigue and minimizing application errors of overlaps and skips. Most light bar systems include a differential GPS (DGPS) receiver and antenna, some kind of computer or microprocessor, and a light bar or graphics display (Stombaugh 2002).

From the study of cost maps, Kayrouz (2008) indicated that inaccuracy of machinery movement, whether in the application stage or the harvesting stage, is very costly. Light bar gives a visual guide that changes colors as the operator veers off the desired path. Previous technologies were developed with the idea of achieving the same goal, for example, planter markers and foam markers. Both of these previous technologies are becoming shadowed by light bar and its effectiveness in improving farm profitability. As a result, foam and planter markers are becoming an obsolete technology (Kayrouz 2008).

2.8.2 Autonomous Field Machinery

According to Haapala et al. (2006), as human capacity to handle simultaneous information is limited, new intelligence has been developed between the user and the system to be controlled. The intelligent layer decides which kind of information is passed on for the user and which part is used for other purposes. Introduction of small, light-weight robots that can perform agricultural tasks autonomously may prove to be a realistic option for farmers in the future. These robots will likely operate in fleets and utilize intelligent controls to cooperate with each other to perform tasks such as scouting for weeds and diseases, yield and field mapping, and plant-specific operations like sowing and fertilizing (Shockley 2010). The role of automation should be designed so that automation helps the individual user to simultaneously avoid stress and increase efficiency of work.

2.9 Conclusions

Precision farming is still only a concept in many developing countries, and strategic support from the public and private sectors is essential to promote its rapid adoption. Successful adoption, however, comprises at least three phases including exploration, analysis, and execution. Precision agriculture can address both economic and environmental issues that surround production agriculture today. Questions remain about cost-effectiveness and the most effective ways to use the technological tools we now have, but the concept of "doing the right thing in the right place at the right time" has a strong intuitive appeal. In the light of today's urgent need, there should be an all-out effort to use new technological inputs to make the "Green Revolution" as an "Evergreen Revolution." Ultimately, the success of precision agriculture depends largely on how well and how quickly the knowledge needed to guide the new technologies can be found.

References

Adamchuck, V. I., & Mulliken, J. (2005). *Site specific management of soil pH (FAQ)*. University of Nebraska-Lincoln, Extension EC05705.

Adrian, A. M., Norwood, S. H., & Mask, P. L. (2005). Producers' perceptions and attitudes toward precision agriculture technologies. *Computers and Electronics in Agriculture, 48*(3), 256–271.

Atherton, B. C., Morgan, M., Shearer, S. A., Stombaugh, T. S., & Ward, A. D. (1999). Site-specific farming: A perspective on information needs, benefits and limitations. *Journal of Soil and Water Conservation, 54*(2), 455–461.

Blackmore, B. S. (2003). *The role of yield maps in precision farming*. Doctoral thesis, Silsoe College, Cranfield University.

Blackmore, B. S., & Marshall, C. J. (1996). Yield mapping; errors and algorithms. In P. C. Robert, R. H. Rust, & W. E. Larson (Eds.), *3rd international conference on precision agriculture* (pp. 403–415). Madison: ASA, CSSA, SSSA & ASAE.

Dillon, C. R., Shearer, S., & Pitla, S. (2007). Improved profitability via enhanced variable rate application management resolution in grain crop production. In J. V. Stafford (Ed.), *Referred proceedings of the 6th European conference on precision agriculture* (pp. 819–825). Skiathos: Wageningen Academic Publishers.

Gandonou, J. M. (2005). *Essays on precision agriculture technology adoption and risk management*. Phd Thesis, University of Kentucky.

Grisso, R., Alley, M., Thomason, W., Holshouser, D., & Roberson, G. T. (2011). *Precision farming tools: Variable-rate application*. Virginia Cooperative Extension, College of Agriculture and Life Sciences, Virginia Polytechnic Institute and State University.

Haapala, H. E. S., Pesonen, L., & Nurkka, P. (2006). Usability as a challenge in precision agriculture – Case study: An ISOBUS VT. Agricultural Engineering International: the CIGR Ejournal. Vol. VIII, Manuscript MES 05 001.

Ima, C. S., & Mann, D. D. (2004). Ergonomic concerns with lightbar guidance displays. *Agricultural Safety and Health, 10*(2), 91–102.

Kayrouz, B. M. (2008). *Precision agriculture: Realizing increased profit and reduced risk through cost map and lightbar adoption*. Master's theses. University of Kentucky.

Khan, A. M. (2013). *Precision farming technique for water resource management* (Doctoral dissertation, Aligarh Muslim University).

Khosla, R. (2008). *The 9th international conference on precision agriculture opening ceremony presentation*. July 20–23rd.

Mickle, E. (2009). *Using GIS to locate areas for growing quality coffee in Honduras*. Environmental studies undergraduate student honors thesis, Paper 3.

Mooney, D. F., Roberts, R. K., Larson, J. A., & English, B. C. (2009). *Economics of the variable rate technology investment decision for agricultural sprayers*. Paper presented at the southern agricultural economics association annual meeting, Atlanta, Georgia.

National Research Council. (1997). *Precision agriculture in the 21st century*. Washington, DC: National Academy Press.

Nemenyi, M., Mesterhazi, P. A., Pecze, Z., & Stepan, Z. (2003). The role of GIS and GPS in precision farming. *Computers and Electronics in Agriculture, 40*(1–3), 45–55.

Olson, K. (1998, August). Precision agriculture: Current economic and environmental issues. In *Sixth Joint Conference on Food, Agriculture, and the Environment* (Vol. 31).

Price, M. (2006). *Mastering ArcGIS* (Vol. 10020, Second ed.). New York: McGraw-Hill.

Robert, P., Rust, R. & Larson, W. (1994). *Sitespecific management for agricultural systems.* Proceedings of the 2nd International Conference on Precision Agriculture, Madison, WI. ASA/CSSA/SSSA.

Roberts, R., English, B., & Mahajanashetti, S. (2001). Environmental and economic effects of spatial variability and weather. In S. Blackmore & G. Grenier (Eds.), *Proceedings of the 3rd European conference on Precision agriculture* (pp. 545–550). Montpellier: AGRO.

Shearer, S. A., Fulton, J. P., McNeill, S. G. & Higgins, S. F. (1999). *Elements of precision agriculture: basics of yield monitor installation and operation.* Cooperative Extension Service, University of Kentucky College of Agriculture.

Shockley, J. M. (2010). *Whole farm modeling of precision agriculture technologies.* University of Kentucky, Doctoral Dissertations, Paper 105.

Stombaugh, T. S. (2002). *Lightbar guidance aids.* Cooperative Extension Service, College of Agriculture, University of Kentucky.

Stombaugh, T., Shearer, S. A., Fulton, J. M., & Ehsani, R. (2002). *Elements of a dynamic GPS test standard.* ASAE Paper Number: 021150. St. Joseph: American Society of Agricultural Engineers.

Swinton, S. M., & Lowenberg-DeBoer, J. (1998). Evaluating the profitability of site-specific farming. *Journal of Production Agriculture, 11*(4), 439–446.

Thom, W. O., Wells, K. L., & Murdock, L. W. (1997). *Taking soil test samples.* Cooperative extension service, College of Agriculture, University of Kentucky.

Chapter 3
Tool and Technologies in Precision Agriculture

Abstract In order to gather and use information effectively, it is important for anyone considering precision farming to be familiar with the technological tools available. These tools include hardware, software, and recommended practices. In this chapter different tools and techniques used in implementation of precision farming have been discussed.

Keywords GPS · Sensors · GIS · Grid soil sampling · VRT · Soil mapping · Precision irrigation · Yield monitor · Precision livestock farming

3.1 Global Positioning System (GPS)

Satellite-based navigation is commonly known as the global positioning system (GPS). The earliest use of GPS dates back to the 1970s and was developed by the US Department of Defense unit. The main purpose for developing GPS was to aid military groups for positioning and timing information. The present day sees diverse application for GPS including commercial use, agriculture, robotics, and clock synchronization.

GPS is a navigation system based on a network of satellites (Fig. 3.1) that helps users to record positional information (latitude, longitude, and elevation) with an accuracy of between 100 and 0.01 m (Lang 1992). GPS allows farmers to locate the exact position of field information, such as soil type, pest occurrence, weed invasion, water holes, boundaries, and obstructions. There is an automatic controlling system, with light or sound guiding panel (DGPS), antenna, and receiver. GPS satellites broadcast signals that allow GPS receivers to calculate their position. The system allows farmers to reliably identify field locations so that inputs (seeds, fertilizers, pesticides, herbicides, and irrigation water) can be applied to an individual field, based on performance criteria and previous input applications (Batte 2000).

Global Positioning System satellites broadcast signals that allow GPS receivers (Fig. 3.2) to calculate their position. This information is provided in real time, meaning that continuous position information is provided while in motion. Having

L. Ahmad, S. S. Mahdi, *Satellite Farming*,
https://doi.org/10.1007/978-3-030-03448-1_3

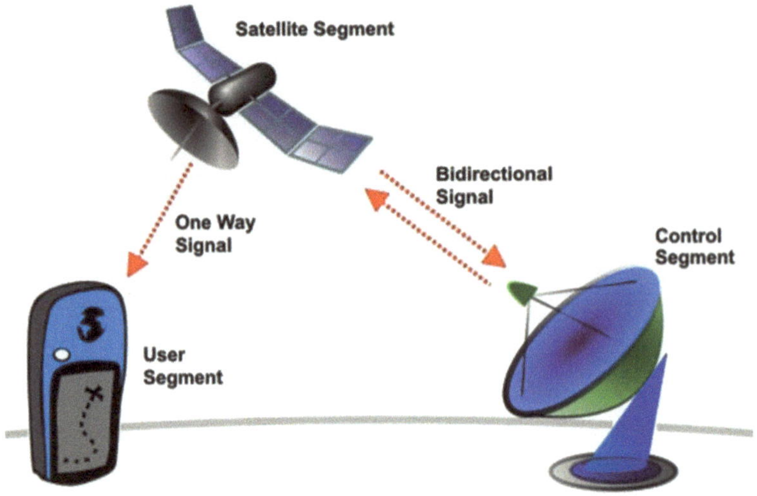

Fig. 3.1 Illustration of the three core segments to a Global Positioning System. (Hinch 2010)

Fig. 3.2 A GPS device

precise location information at any time allows soil and crop measurements to be mapped. GPS receivers, either carried to the field or mounted on implements, allow users to return to specific locations to sample or treat those areas. Uncorrected GPS signals have an accuracy of about 300 feet.

3.2 Sensor Technologies

Various technologies such as electromagnetism, conductivity, photoelectricity, and ultrasound are used to measure humidity, vegetation, temperature, texture, structure, physical character, humidity, nutrient level, vapor, air, etc. Remote sensing data are used to distinguish crop species, locate stress conditions, identify pests and weeds, and monitor drought, soil, and plant conditions. Sensors enable the collection of immense quantities of data without laboratory analysis (Chen et al. 1997). Some of the commonly used sensors in precision agriculture are given in Table 3.1.

Table 3.1 Commonly used sensors in precision agriculture

Sensor	Applications
Temperature sensor	Used to measure temperature with an electrical output proportional to the temperature (in °C). It can measure temperature more accurately than a using a thermostat (Fig. 3.3)
Humidity sensor	Used to measure the presence of water in land. The humidity sensor has its output proportional to the temperature (in RH %). The operating range is for humidity is generally 20–95% RH and for temperature is 0–60 °C (Fig. 3.3)
See Fig. 3.3	
Water level sensor	Also known as float balls, are spherical, cylindrical, or similarly shaped objects, made from either rigid or flexible material, that are buoyant in water and other liquids. They are non-electrical hardware frequently used as visual sight indicators for surface demarcation and level (Fig. 3.4)
Soil sensor	Provide continual, consistent accuracy measuring a number of parameters including the three most significant soil parameters simultaneously – moisture, salinity, and temperature (Fig. 3.5)
See Fig. 3.4	See Fig. 3.5
LIDAR	Light detection and ranging is used to obtain dynamic measurements to estimate fruit-tree leaf area and combined with GPS has been applied for 3-D map generation in vine plantations. Laser and hyperspectral data are used for tree classification, including coniferous and deciduous trees. 3-D modeling of tomato canopies is obtained through high-resolution portable scanning LIDAR (Fig. 3.6)
See Fig. 3.6	
Capacitance probes	Suitable for measurement of soil moisture in tropical areas
Reflectometers	Can determine moisture content in oil palm fruits
Ultrasonic ranging sensors	Used for analyzing apple tree canopies
Optoelectronic sensors	Weed detection in wide row crops has been analyzed in terms of accuracy and feasibility
Fluorescence-based optical sensor	Used to monitor grape maturation by specifically monitoring anthocyanin accumulation
The infrared thermocouple	Measurement of the leaf temperature of forests or agricultural plants (Fig. 3.7)

(continued)

Table 3.1 (continued)

Sensor	Applications
See Fig. 3.7	See Fig. 3.8
An FPGA-based fused smart sensor	Real-time plant transpiration monitoring is possible
Automated point dendrometers	Analyze tropical tree line stem growth has been applied in forest inventories (Fig. 3.8)
Optical and microwave sensors	Suitable for characterizing olive grove canopies
Optoelectronic sensors	Weed detection in wide row crops has been analyzed in terms of accuracy and feasibility
pH soil-based sensors	Allow measurements of variables in the soil oriented toward crop productivity
See (Fig. 3.9)	
Eddy covariance sensors	Applied to quantify carbon metabolism of peatlands and also regional and global analysis of observations from micrometeorological tower sites
Near-infrared (NIR) sensors	Measure the amount of light reflected back at them at specific wave lengths in order to make this estimate. Estimate the amount of biomass present and calculate future nitrogen requirements
	Can be integrated into systems to control fertilizer application at any point in the field
	Crop scouting prior to fertilizer application to determine the optimum rate to be applied over the whole field

Fig. 3.3 Humidity and temperature sensor on a California farm

Fig. 3.4 Water level sensor

Fig. 3.5 Soil sensor

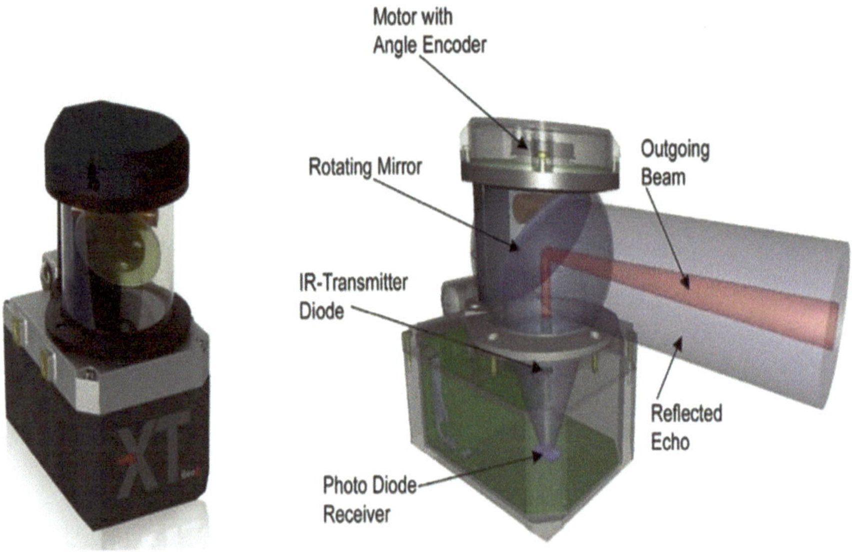

Fig. 3.6 LIDAR

Fig. 3.7 Infrared thermocouple

3.3 Geographic Information System (GIS)

This system comprises hardware, software, and procedures designed to support the compilation, storage, retrieval, and analysis of feature attributes and location data to produce maps. GIS links information in one place so that it can be extrapolated

Fig. 3.8 Automated point dendrometer

when needed. Computerized GIS maps are different from conventional maps and contain various layers of information (e.g., yield, soil survey maps, rainfall, crops, soil nutrient levels, and pests). GIS is a kind of computerized map, but its real role is using statistics and spatial methods to analyze characters and geography. A farming GIS database can provide information on filed topography, soil types, surface drainage, subsurface drainage, soil testing, irrigation, chemical application rates, and crop yield. Once analyzed, this information is used to understand the relationships between the various elements affecting a crop on a specific site. In addition to data storage and display, the GIS can be used to evaluate present and alternative management by combining and manipulating data layers to produce an analysis of management scenarios.

3.4 Grid Soil Sampling and Variable Rate Fertilizer (VRT) Application

Variable rate technologies (VRT) are automatic and may be applied to numerous farming operations. VRT systems set the rate of delivery of farm inputs depending on the soil type noted in a soil map. Information extrapolated from the GIS can control processes, such as seeding, fertilizer and pesticide application, herbicide selection, and application at a variable rate in the right place at the right time. VRT is perhaps the most widely used PFS technology in the United States. Grid soil sampling uses the same principles of soil sampling but increases the intensity of sampling. Soil samples collected in a systematic grid also have location information that allows the data to be mapped. The goal of grid soil sampling is a map of nutrient

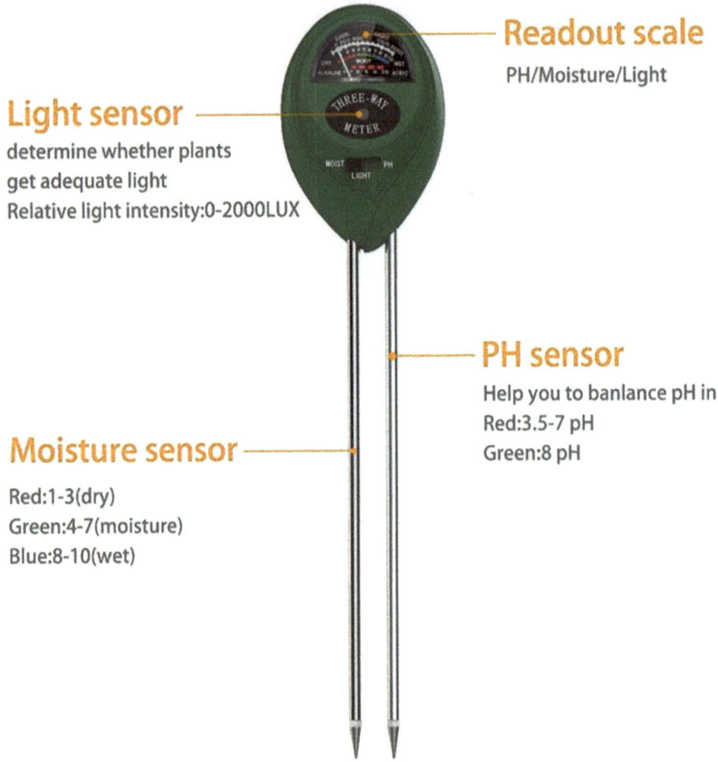

Fig. 3.9 pH meter with sensor

needs, called an application map. Samples may be collected for more than one area of a field which fall into the same range of yield, soil color, etc. and thus the same zone. Grid soil samples are analyzed in the laboratory, and an interpretation of crop nutrient needs is made for each soil sample. Then the fertilizer application map is plotted using the entire set of soil samples. The application map is loaded into a computer mounted on a variable rate fertilizer spreader. The computer uses the application map and a GPS receiver to direct a product delivery controller that changes the amount and/or kind of fertilizer product, according to the application map (Berntsen et al. 2006).

3.5 Digital Soil Mapping

Digital soil mapping is the creation and the population of a geographically referenced soil database, generated at a given resolution by using field and laboratory observation methods coupled with environmental data through quantitative relationships.

There has been some effort to create digital soil maps (DSMs) that combine attributes like soil type, weather pattern, and chemical composition in a GIS to demonstrate the overall fertility of an agricultural region or to predict its production. The most notable example is the regional soil map created by the Natural Resources Conservation Service (NRCS).

3.6 Crop Management

Satellite data provide farmers a better understanding of the variation in soil conditions and topography that influence crop performance within the field. Farmers can, therefore, precisely manage production factors, such as seeds, fertilizers, pesticides, herbicides, and water control, to increase yield and efficiency.

3.7 Rate Controllers

Rate controllers are devices designed to control the delivery rate of chemical inputs such as fertilizers and pesticides, either liquid or granular (Fig. 3.10). These rate controllers monitor the speed of the tractor/sprayer traveling across the field, as well as the flow rate and pressure (if liquid) of the material, making delivery adjustments in real time to apply a target rate. Rate controllers have been available for some time and are frequently used as stand-alone systems.

Fig. 3.10 Display monitor of a rate controller

3.8 Precision Irrigation in Pressurized Systems

Recent developments are being released for commercial use in sprinkler irrigation by controlling the irrigation machines motion with GPS-based controllers. In addition to motion control, wireless communication and sensor technologies are being developed to monitor soil and ambient conditions, along with operation parameters of the irrigation machines (i.e., flow and pressure) to achieve higher water application efficiency and utilization by the crop. These technologies show remarkable potential, but further development is needed before they become commercially available.

3.9 Software

Applying precision agriculture technologies will frequently require the use of software to carry out diverse tasks such as display–controller interfacing, information layers mapping, pre- and post-processing data analysis and interpretation, farm accounting of inputs per field, and many others. The most common are software to generate maps (e.g., yield, soil), software to filtering collected data, software to generate variable rate applications maps (e.g., for fertilizer, lime, chemicals), software to overlay different maps, and software to provide advanced geostatistical features.

All are excellent options for precision agriculture farm management and record keeping to keep up with the needs of modern, information-intensive farming systems. There are a few companies that operate worldwide and provide integrated software packages from generating all different types of maps, having statistical analysis tools and also record keeping. The machinery companies that provide yield meters also offer software to generate yield maps and fertilizer companies provide software to generate variable rate applications maps. Some of the packages are very complicated for farmers to use, and they are fairly expensive, while some others are considerably simpler and cheaper with fewer options. The packages are more user-friendly still problems related to data transfer between farmers and between farmer, coop, and consultant. To overlay maps, mainly soil and yield maps, is also a difficult task so far.

3.10 Yield Monitor

Yield monitors (Fig. 3.11) are a combination of several components. They typically include several different sensors and other components, including a data storage device, user interface (display and key pad), and a task computer located in the combine cab, which controls the integration and interaction of these components. The sensors measure the mass or the volume of grain flow (grain flow sensors), separator speed, ground speed, and grain. In the case of grains, yield is continuously

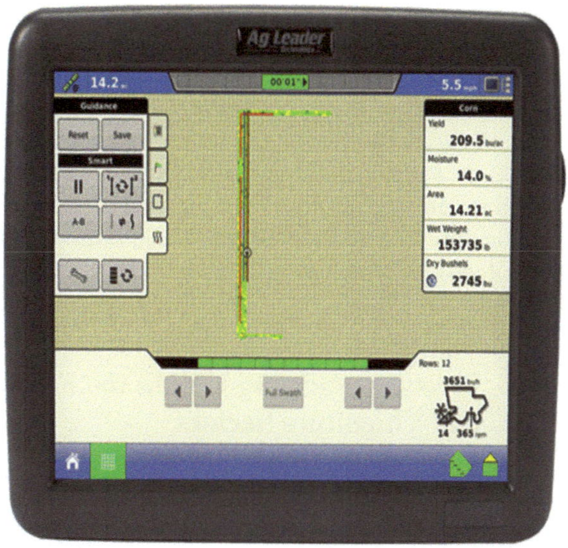

Fig. 3.11 Display console of a yield monitor

recorded by measuring the force of the grain flow as it impacts a sensible plate in the clean grain elevator of the combine.

A recent development of a mass flow sensor works on the principle of transmitting beams of microwave energy and measuring the portion of that energy that bounces back after hitting the stream of seeds flowing through the chutes. In all yield monitors, GPS receivers are used to record the location of yield data and create yield maps. Other yield-monitoring systems include devices used in forage crops to keep track of weight, moisture, and other information on a per bale basis (Davis et al. 1998).

3.11 Precision Farming on Arable Land

The use of PA techniques on arable land is the most widely used and most advanced among farmers (Bowman 2008). CTF is a whole farm approach that aims at avoiding unnecessary crop damage and soil compaction by heavy machinery, reducing costs imposed by standard methods. Controlled traffic methods involve confining all field vehicles to the minimal area of permanent traffic lanes with the aid of GNSS technology and decision support systems.

Another important application of precision agriculture in arable land is to optimize the use of fertilizers, starting with the three main nutrients nitrogen, phosphorus, and potassium. In conventional farming, these fertilizers are applied uniformly over fields at certain times during the year. This leads to over-application in some places and under-application in others. The environmental cost is directly related to over-application which allows nitrogen and phosphorus leaching from the field into ground- and surface waters or to other areas of the field where they are not desired.

With the use of precision agriculture methods, fertilizers can be applied in more precise amounts, with a spatial and temporal component to optimize the application. The technology that allows the farmer to control the amount of inputs in arable lands is the variable rate application (VRA), which combines a variable rate (VR) control system with application equipment to apply inputs at a precise time and/or location to achieve site-specific application rates of inputs. VRs are decided on the basis of prior measurement, e.g., from remote sensing or machine mounted.

3.12 Precision Farming Within the Fruits and Vegetables and Viticulture Sectors

In fruit and vegetable farming, the recent rapid adoption of machine vision methods allows growers to grade products and to monitor food quality and safety, with automation systems recording parameters related to product quality. These include color, size, shape, external defects, sugar content, acidity, and other internal qualities (Njoroge et al. 2002). Additionally, tracking of field operations such as chemicals sprayed and use of fertilizers can be possible to provide complete fruit and vegetable processing methods.

This information can be disclosed to consumers for risk management and for food traceability as well as to producers for precision agriculture to get higher quality and larger yields with optimized inputs. In recent years, several new approaches were developed that take into account the actual size of the tree, the condition of the crop, but also the environmental conditions (Doruchowski et al. 2009).

The development and adoption of PA technologies and methodologies in viticulture (termed precision viticulture, PV) are more recent than in arable land. However, driven by the high value of the crop and the importance of quality, several research projects already exist in wine production areas of the world (Ferreiro-Arman et al. 2006). Grape quality and yield maps are of great importance during harvest to avoid mixing grapes of different potential wine qualities. The parcels with greatest opportunities for PV are those which reveal a high degree of yield variation. A high degree of variation will mean higher VRA of inputs and, therefore, greater economic and environmental benefit in comparison with uniform management.

3.13 Precision Livestock Farming (PLF)

Precision livestock farming (PLF) is defined as the management of livestock production using the principles and technology from precision agriculture. Processes suitable for the precision livestock farming approach include animal growth, milk and egg production, detection and monitoring of diseases and aspects related to animal behavior, and the physical environment such as the thermal microenvironment and emissions of gaseous pollutants. Systems include milk monitoring to check fat and microbial levels, helping to indicate potential infections, as well as new robotic feeding systems, weighing systems, robotic cleaners, feed pushers, and other aids for the stockman such as imaging systems to avoid direct contact with animals. New systems for data monitoring for feed and water consumption can be used to the early detection of infections is available now (Fig. 3.12).

Other developments include the monitoring on the growing herd where measurement of growth in real time is important to provide producers with feed conversion and growth rates. Acoustic sensors detect an increase in coughing of pigs as an indicator of respiratory infection. Other sensors are now used to provide alerts concerning birthing and fertility.

A vaginal thermometer monitors the temperature, the imminence of birthing, and the breaking of waters and communicates to the farmer via SMS. Also, a sensor placed on an animal's collar records parameters to detect signs of oestrus and the readiness for fertilization. An SMS message then allows the farmer to plan for insemination.

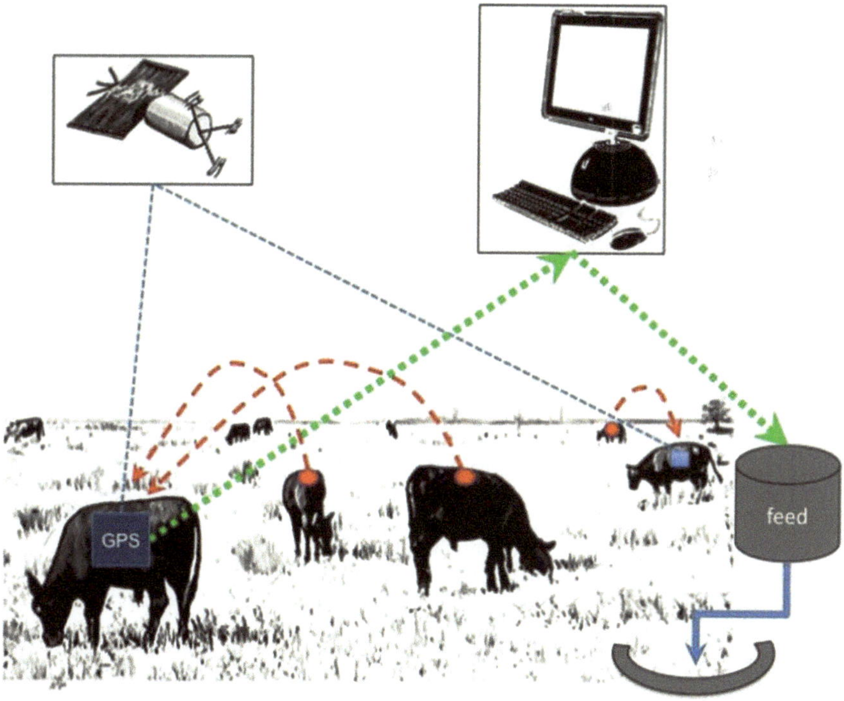

Fig. 3.12 Precision livestock farming

3.14 Online Resources for Precision Agriculture

There is a wealth of information available over the Internet on new technology for farm production. Most manufacturers of farm equipment, GPS receivers, sensors, and other PA technologies use this media to inform growers on new products, technical specifications, troubleshooting information, software upgrades, and a variety of services.

3.15 Conclusions

Precision farming is still only a concept in many developing countries, and strategic support from the public and private sectors is essential to promote its rapid adoption. Successful adoption, however, comprises at least three phases including exploration, analysis, and execution. Precision agriculture can address both economic and environmental issues that surround production agriculture today. Questions remain about cost-effectiveness and the most effective ways to use the technological tools we now have, but the concept of "doing the right thing in the right place at the right time" has a strong intuitive appeal. In the light of today's urgent need, there should be an all-out effort to use new technological inputs to make the "Green Revolution" as an "Evergreen Revolution." Ultimately, the success of precision agriculture depends largely on how well and how quickly the knowledge needed to guide the new technologies can be found.

Precision farming provides a new solution using a systems approach for today's agricultural issues such as the need to balance productivity with environmental concerns. It is based on advanced information technology. It includes describing and modeling variation in soils and plant species and integrating agricultural practices to meet site-specific requirements. It aims at increased economic returns, as well as at reducing the energy input and the environmental impact of agriculture.

References

Batte, M. T. (2000). Factors influencing the profitability of precision farming systems. *Journal of Soil and Water Conservation, 55*(1), 12–18.

Berntsen, J., Thomsen, A., Schelde, K., Hansen, O. M., Knudsen, L., Broge, N., Hougaard, H., & Hørfarter, R. (2006). Algorithms for sensor-based redistribution of nitrogen fertilizer in winter wheat. *Precision Agriculture, 7*(2), 65–83.

Bowman, K. (2008, August). Economic and environmental analysis of converting to controlled traffic farming. In *6th Australian controlled traffic farming conference* (pp. 61–68).

Chen, F., Kissel, D. E., Clark, R., West, L. T., Rickman, D., Luval, J., & Adkin, W. (1997). *Determining surface soil clay concentration at a field scale for precision agriculture*. Huntsville: University of Georgia.

Davis, G., Casady, W. W., & Massey, R. E. (1998). *Precision agriculture: An introduction*. Columbia: Extension Publications (MU).

Doruchowski, G., Basari, P., & van de Zande, J. C. (2009, January 05–09). Precise spray application in fruit growing according to crop health status, target characteristics and environmental circumstances. In *Proceedings of the 8th fruit, nut and vegetable production engineering symposium, Concepción, Chile* (pp. 494–502).

Ferreiro-Arman, M., Da Costa, J. P., Homayouni, S., & Martin-Herrero, J. (2006, September). Hyperspectral image analysis for precision viticulture. In *International conference image analysis and recognition* (pp. 730–741). Berlin/Heidelberg: Springer.

Hinch, S. W. (2010). *Outdoor navigation with GPS*. Birmingham: Wilderness Press, Keen Communication.

Lang, L. (1992). GPS, GIS, remote sensing: An overview. *Earth Observation Magazine*, 23–26.

Njoroge, J. B., Ninomiya, K., Kondo, N., & Toita, H. (2002, August). Automated fruit grading system using image processing. In *SICE 2002. Proceedings of the 41st SICE annual conference*, *2*, 1346–1351. IEEE.

Chapter 4
Precision Soil Sampling and Tillage

Abstract Different methods of soil sampling used for precision nutrient and soil management have been put forth. In this chapter, the reader can also find a description of types and importance of tillage in agriculture. A brief description of different terms associated with tillage is also included. Finally, modern concepts in tillage including precision tillage have also been discussed.

Keywords Grid soil sampling · Zone sampling · Prescription maps · Primary tillage · Secondary tillage · Minimum tillage · Zero tillage · Precision tillage

4.1 Introduction

Site-specific nutrient management depends on a sound inventory of soil nutrient availability. Site-specific nutrient management for crop production begins with an inventory of soil test nutrient levels in a field. Fertilizer recommendations are based on expected response to fertilizer application as a function of soil test levels. Therefore, site-specific fertilizer applications can be no better than the accuracy of the soil test map from which the fertilizer recommendations are based. Precision usually increases as fields are divided and sampled as smaller areas (Fig. 4.1).

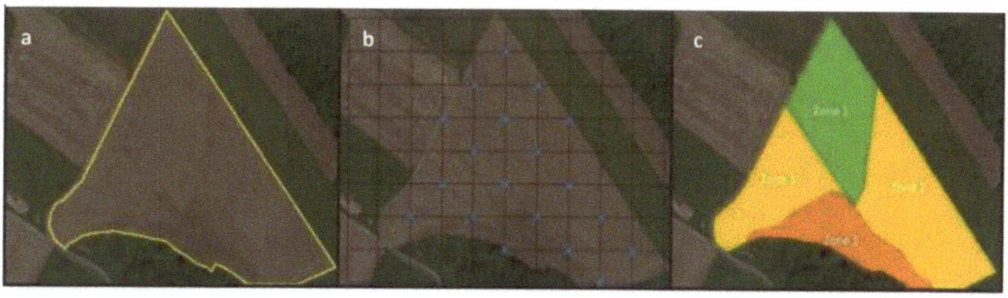

Fig. 4.1 Soil samples may be collected for the (**a**) whole field, (**b**) on a consistently spaced grid or (**c**) by management zones

L. Ahmad, S. S. Mahdi, *Satellite Farming*,
https://doi.org/10.1007/978-3-030-03448-1_4

4.2 Soil Sampling

Less than a teaspoonful of soil is actually used for the laboratory analysis. That small amount must represent the entire area for which the recommendation is to be made. For example, in a traditional sampling scheme, one teaspoon or less of soil represents up to 40 acres (i.e., more than 80 million pounds of soil in the top 7 inches).

With site-specific management being implemented on many farms, there is a growing need to characterize the variability in nutrient needs across the field. Each sample should represent the field for best characterization of the variability within the field, to serve as a guide for variable rate application of crop nutrients. Where field variability is low, larger sample areas are acceptable; where variability is high, more samples are needed to adequately represent the field.

4.3 Sampling Procedures

The goal of soil sampling is to estimate the capacity of the soil to provide adequate amounts of the necessary nutrients to meet the needs of the crop (or crops) to be grown. It should be clearly understood that soil testing does not measure the amount of nutrients in the soil. The test results can only be used in conjunction with a calibration curve that relates the laboratory analysis to a set of crop response data. Without the response (calibration) data, the laboratory results are meaningless. The samples should be collected in such a way as to best meet that goal. The sampling pattern should be set to best characterize the variability within the field.

4.3.1 Depth

Before sampling, check with the laboratory that will conduct the analysis to see what sampling depth is recommended. Sampling depth should be determined to represent the root zone the plant will draw from, but should also be consistent with the sampling depth used in developing the calibration data set to be used for interpreting the soil tests. Most soil test calibrations are based upon a 6- to 8-inch depth, most commonly 6 2/3 inches. In dry years, when it is difficult to push the probe into the ground, there is a danger of not getting the proper depth. Sampling too shallow will often lead to unusually high soil test results, because of the tendency for nutrients to become concentrated near the surface. Shallow sampling will thus overestimate the actual soil nutrient status and lead to underestimating fertilizer rates needed. This problem is even greater in reduced-tillage systems. Uniformity of soil sampling depth is one of the most critical parts of soil testing, yet it is one of the most common sources of error.

4.3.2 Pattern

Whether variable rate nutrient application is planned or not, sampling the soil in an organized pattern is a good management practice. It helps ensure adequate representation of the entire field. Most agronomists recommend sampling on a pattern so that each sample represents about 2 ½ acres (1 ha) or less. At least one sample per acre is preferred, especially in areas receiving 25″ or more of annual rainfall and in irrigated fields.

4.4 Soil Sampling Instructions

Accurate soil analysis with meaningful interpretation requires properly taken samples. Follow all directions carefully and correctly. Sampling technique presents the greatest chance for errors in results. Laboratory analytic work will not improve the accuracy of a sample that does not represent the area.

(i) *Select the Proper Equipment*

Collect samples using chrome plated or stainless steel sampling tubes or augers. Avoid galvanized, bronze, or brass tools. Use clean, plastic buckets. Do not use galvanized or rubber buckets, as they will contaminate the samples.

(ii) *When to Take Samples*

Sampling can take place during any period of the year. However, it is best to sample a field at about the same time of year. Wait a minimum of 30 days to sample after applications of fertilizer, lime, or sulfur.

(iii) *Sample Area*

Samples must be representative of the area you are treating. Most often, sampling by soil color is an acceptable method for dividing large fields into "like" areas. County ASCS aerial photographs can be used as a guide. Areas that differ in slope, drainage, past treatment, etc. should be sampled separately. Sampling across dissimilar soil types is not recommended. And finally, the sample area should be large enough for special lime or fertilizer treatments.

Always remember to remove any surface debris prior to sampling.

Do not sample:

- Dead or back furrows
- Fence rows, old or new
- Old roadbeds or near limestone gravel roads
- Terrace channels
- Wind breaks or snow fence lines
- Turn-rows

- Spill areas
- Fertilizer bands including anhydrous N
- Unusual or abnormal spots

(iv) *Sample Depth*

Sampling depth must remain consistent because many soils are stratified, and variation in depth will introduce errors into the analytical results. To test for soil stratification, sample through the soil profile, separately, 0" to 2", 2" to 4", 4" to 6", and 6" to 8". Remember to take the recommended number of cores per sample. The greater the difference in the analytical data between samples, the greater the degree of stratification.

(v) *Numbers of Cores and Acres per Sample*

Various studies have shown that proper sampling requires at least 10 cores per sample, and sometimes 15 or more cores, depending on the nature of the soil and size of the area being sampled. A smaller number can introduce variability into the results from different sampling years. There is no rule for the number of acres to include in a single sample. This must depend on the local situation. However, the University of Illinois has long recommended that a single sample should represent no more than 5 acres.

(vi) *Preparing Samples for Shipment*

Thoroughly mix the randomly taken core samples in a plastic bucket, and remove a well-mixed composite sample (1/2 to 1 pint) from the mixture. Place it into the lab's sample bag, filling it to the "line." All samples taken for nitrogen analyses should be immediately air-dried, shipped early in the week, or shipped frozen.

Once the sample is in the bag, fold the top down to exclude air, roll it down to close, and fold the tabs. Write your sample ID designation (include grid subsample identification where applicable) and your customer's name on the bag where requested. Complete all the remaining information as required.

4.5 Sampling Pattern Options

The sampling pattern should be selected to best represent the field, accounting for known sources of variability (major soil-type changes, past cropping patterns, etc.). A grid pattern is usually the best way to be sure the entire field is represented, but with the possibility of patterns developing from past nutrient applications, cropping effects, and other uniform patterns, it is advisable to use a sampling scheme that avoids arranging sampling points in a straight line.

4.6 Grid Soil Sampling

The common approach to achieve systematic soil sampling is to overlay a square or rectangular grid on a map or photograph of the field, identify and drive to the middle of each grid cell, and collect a soil sample at that point (Fig. 4.2). The soil sample consists of several soil cores collected within a small radius of the cell center (Clay et al. 1997, 2002). The soil cores are composited and bagged as one soil sample for analysis at a soil testing laboratory. The purpose of compositing several cores is to average or "bulk" out variability in soil test properties that occurs over small distances (Wollenhaupt et al. 1994).

Grid cell sampling can be efficiently conducted by counting crop rows and using distance measuring devices to locate sampling points. While easy to implement in the field, this practice can lead to bias. Tillage, fertilizer application, drainage, old field boundaries, and cropping patterns tend to occur in regular patterns across fields. If the grid sampling pattern is a multiple or fraction of other patterns, the soil samples may not correctly represent the soil test variability within the field.

The potential for bias can be minimized by shifting the sampling locations to the right or left of the cell center in alternating rows perpendicular to the management pattern (e.g., row direction). The resulting sampling grid takes on the appearance of a diamond pattern (Fig. 4.3). This sampling pattern can also be implemented by counting rows and measuring distances.

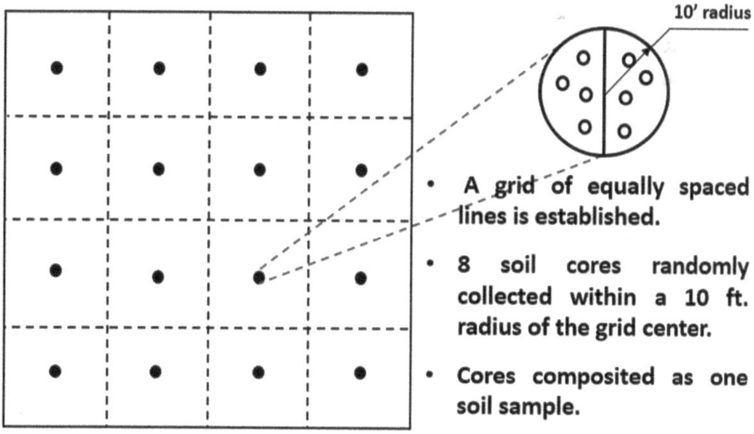

Fig. 4.2 Schematic showing the layout of a square grid and locations where soil cores would be collected

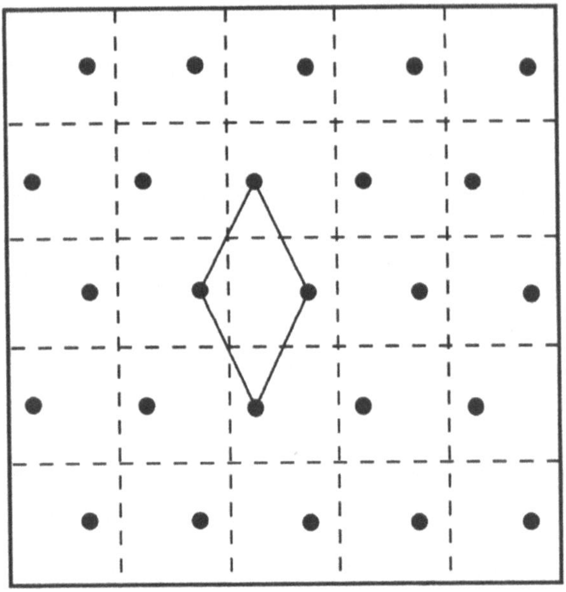

Fig. 4.3 Modification of a square grid where alternating rows of sample points are shifted one half the distance from the cell center and edge

4.6.1 Advantages and Disadvantages of Grid Soil Sampling

The advantages and disadvantages of the method are listed below:

Advantages

- Assess nutrient variability present in field
- Identify hot spots/trouble areas
- Minimize excess nutrient application
- Target inputs where needed
- Minimum skill level

Disadvantages

- No justification for grid sizes
- Grid arbitrarily placed in field
- Ignores soil properties and field characteristics
- Labor and time intensive
- Expensive

4.7 Zone Sampling

Zone sampling involves dividing the field into zones that are uniform enough to be managed as a whole and then sampling to determine the average soil test values for those zones. The success of the zone sampling relies on the amount and quality of the data used to determine the zones. Layers such as soils maps, aerial photos, yield maps, topographic maps, management history, and personal field experience can provide valuable information about the variation in a field. This information can be used to define sample zones or management zones in a field. As the number of management zones in a field increases, the number of samples needed increases. If only a few zones exist, samples can be combined to reduce the cost of analytical expenses (Hawkins et al. 2016).

The management zone is better choice than grids when the operator has a long history of working with the field and topography varies and can be used to define zones and where yield map data over time has defined high- and low-yielding areas, the soil-type map represents yield zones, or other remote sensing data is available to overlay with operator experience to define yield patterns in a field.

It is important to note that differences in yield may not be always be caused by differences in soil test values. Identifying other yield-limiting factors will help fine tune your soil fertility program for each field.

4.7.1 Sampling Method

The goal when sampling by management zone is to determine the best estimate of the entire zone. If the data used to determine the zones is accurate, the soil test values should be relatively consistent. In this case, taking multiple soil cores is necessary to reduce the chance of pulling one from a "bad spot."

Figure 4.4 shows the recommended pattern for pulling soil cores when zone sampling. Sample points should be taken randomly (recommended to walk in a zigzag pattern) with 10–15 cores per sample area up to 25 acres. Georeferenced sample points may give a better opportunity to compare sample trends over time by returning to near the same point in future years. This can be beneficial to tracking soil fertility recommendation program effects on soil test levels over time.

Since the soil test values will represent the average for the entire zone, interpolation should not be used. A blanket fertilizer application rate within each zone is most appropriate when zone sampling is used.

4.7.2 Advantages and Disadvantages of Zone Sampling

The advantages and disadvantages of the method are listed below:

Advantages

- Zones are delineated based on past field performance and intrinsic properties of the field.
- The method classifies spatial variability.
- Requires less time and labor than grid sampling method.
- More economical than grid sampling method.

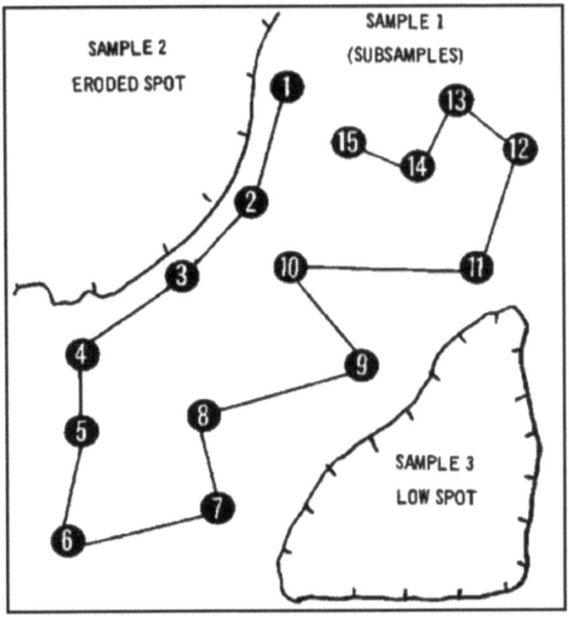

Fig. 4.4 Recommended pattern for collecting soil cores for zone sampling

Disadvantages

- Greater initial time and financial investment to implement zone management
- Higher skill level required
- Requires field knowledge

4.8 Prescription Maps

A prescription map is an electronic data file containing specific information about input rates to be applied in every zone of a field. Prescriptions are not pictures; they are files. Rates are selected for each zone. These rates are built into a prescription file, basically a GPS coordinate with a fertilizer and/or seeding rate associated with it. Prescription files are specific to the type of controller used. Prescriptions can be written for any controller. This file is loaded on to the controller and the rates change as you move across the field.

A prescription map tells the controller how much product to apply based on the location of the equipment in the field. One must know what computer and/or controller will be used to create a prescription map in the appropriate format. Most agricultural GIS packages can create prescription maps in multiple formats. Other factors to note are equipment limitations (minimum and maximum rates) and application width. Some controllers require that a boundary file of the field be included in addition to the prescription map. The prescription is written to a compact flash

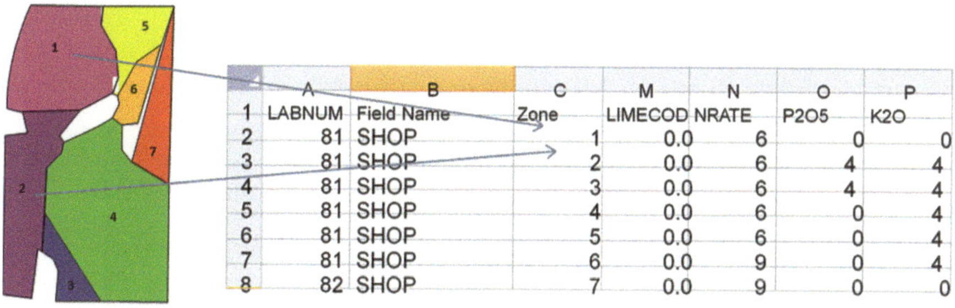

	A	B	C	M	N	O	P
1	LABNUM	Field Name	Zone	LIMECOD	NRATE	P2O5	K2O
2	81	SHOP	1	0.0	6	0	0
3	81	SHOP	2	0.0	6	4	4
4	81	SHOP	3	0.0	6	4	4
5	81	SHOP	4	0.0	6	0	4
6	81	SHOP	5	0.0	6	0	4
7	81	SHOP	6	0.0	9	0	4
8	82	SHOP	7	0.0	9	0	0

Fig. 4.5 A prescription map with seven defined zones. The spreadsheet contains soil test data indicating how zones are linked to soil test results

(CF), PCMCIA card (depending on equipment selections), or other type of data card, which is then uploaded to the computer within the machine cab.

For prescriptions to be based on zones, the first step is to delineate the zones. This step can be completed using several methods. The spatial form can be imported into the GIS packages and clipped to field boundaries. Farmer knowledge and yield data are also excellent tools for creating management zones. When zone maps are created, each zone within the field is automatically assigned a number (Fig. 4.5). The soil sample is usually named in the following format: Farm Name-Field Name-Zone Name. When soil analysis data is returned from the lab, the same naming convention is used, which allows the GIS to match the data to the appropriate zone (Fig. 4.5). Zones can be used for numerous applications, such as variable rate fertilizer (broadcast and sidedress), plant growth regulators, cotton defoliants, and seeding rates. For grid sampling, the boundary is typically divided into 2 ½ acre grids, using a desktop GIS or a PDA (handheld computer) loaded with an agricultural GIS software package. Grids are usually arbitrary and do not account for any past or current knowledge of the field. Each grid assigned a number, sometimes referred to as Feature ID. Feature ID is comparable to the "zone" column in Fig. 4.5; this is how the GIS package links the soil test results back to the geographical location.

4.9 Tillage

Tillage refers to any manipulation of the soil aimed at providing conditions necessary for crop growth. In a classical modern production system, a machine is used to manipulate the soil in order to provide the conditions required for the crop to grow. In this chapter, the reader can find a description of types and importance of tillage in agriculture. A brief description of different terms associated with tillage is also included. Finally, modern concepts in tillage including precision tillage have also been discussed.

These conditions vary according to the need including weed control, providing desired soil structure, incorporation of residues, preparation of land for irrigation, mixing fertilizer and other soil amendments into the soil, and destroying insects, pest, etc. (Anazodo et al. 1991). In the past farmers used plankers which were drawn by animals and small tractors; despite labor-intensive practices, accuracy and precious results were not achieved.

Tilth is the physical condition of soil obtained out of tillage, or it is the result of tillage. The tilth may be a coarse tilth, fine tilth, or moderate tilth.

4.10 Objectives of Tillage

The main objectives of tillage are:

(i) To prepare a good seedbed which helps the germination of seeds.
(ii) To create conditions in the soil suited for better growth of crops.
(iii) To control the weeds effectively.
(iv) To make the soil capable for absorbing more rainwater.
(v) To mix up the manure and fertilizers uniformly in the soil.
(vi) To aerate the soil.
(vii) To provide adequate seed–soil contact to permit water flow to seed and seedling roots.
(viii) To remove the hardpan and to increase the soil depth.
(ix) To achieve these objectives, the soil is disturbed/opened up and turned over.

4.11 Terms Associated with Tillage

4.11.1 Types of Tillage

Tillage operations may be grouped into:

1. On-season tillage
2. Off-season tillage
3. Special-purpose tillage

1. **On-Season Tillage**

Tillage operations that are done for raising crops in the same season or at the onset of the crop season are known as on-season tillage. They may be preparatory cultivation and after cultivation.

Preparatory Tillage This refers to tillage operations that are done to prepare the field for raising crops. It consists of deep opening and loosening of the soil to bring about a desirable tilth as well as to incorporate or uproot weeds and crop stubble when the soil is in a workable condition.

Types of preparatory tillage:

(a) Primary tillage
(b) Secondary tillage

(a) *Primary tillage*: The tillage operation that is done after the harvest of crop to bring the land under cultivation is known as primary tillage or ploughing. Ploughing is the opening of compact soil with the help of different ploughs. Country plough, moldboard plough, bose plough, tractor, and power tiller drawn implements are used for primary tillage (Fig. 4.6).

(b)*Secondary tillage*: The tillage operations that are performed on the soil after primary tillage to bring a good soil tilth are known as secondary tillage. Secondary tillage consists of lighter or finer operation which is done to clean the soil, break the clods, and incorporate the manure and fertilizers. Harrowing and planking are done to serve those purposes.

Planking is done to crush the hard clods, to level the soil surface, and to compact the soil lightly. Harrows, cultivators, Guntakas, and spade are used for secondary tillage (Fig. 4.7).

(c)*Layout of seedbed*: This is also one of the components of preparatory tillage. Leveling board, buck scrapers, etc. are used for leveling, and markers are used for layout of seedbed.

After Cultivation (Intertillage) The tillage operations that are carried out in the standing crop after the sowing or planting and prior to the harvesting of the crop

Fig. 4.6 Primary tillage operation using disc plough

Fig. 4.7 Secondary tillage using a cultivator

plants are called after tillage. This is also called as inter-cultivation or post-seeding/planting cultivation. It includes harrowing, hoeing, weeding, earthing up, drilling, or side-dressing of fertilizers, etc. Spade, hoe, weeders, etc. are used for inter-cultivation.

2. **Off-Season Tillage**

Tillage operations done for conditioning the soil suitably for the forthcoming main season crop are called off-season tillage. Off-season tillage may be postharvest tillage, summer tillage, winter tillage, and fallow tillage.

3. **Special-Purpose Tillage**

Tillage operations intended to serve special purposes are said to be special-purpose tillage. They are:

(a) *Subsoiling:* To break the hardpan beneath the plough layer, special tillage operation (chiseling) is performed to reduce compaction. Subsoiling is essential and once in 4–5 years where heavy machineries are used for field operations, seeding, harvesting, and transporting. Advantages of subsoiling are greater volume of soil may be obtained for cultivation of crops, excess water may percolate downward to recharge the permanent water table and reduce runoff and soil erosion, and roots of crop plants can penetrate deeper to extract moisture from the water table.

(b) *Clean tillage:* It refers to working of the soil of the entire field in such a way no living plant is left undisturbed. It is practiced to control weeds, soilborne pathogen, and pests.
(c) *Blind tillage:* It refers to tillage done after seeding or planting the crop (in a sterile soil) either at the preemergence stage of the crop plants, or while they are in the early stages of growth so that crop plants (sugarcane, potato, etc.) do not get damaged, but extra plants and broad-leaved weeds are uprooted.
(d) *Dry tillage:* Dry tillage is practiced for crops that are sown or planted in dry land condition having sufficient moisture for germination of seeds. This is suitable for crops like broadcasted rice, jute, wheat, oilseed crops, pulses, potato, and vegetable crops. Dry tillage is done in a soil having sufficient moisture (21–23%). The soil becomes more porous and soft due to dry tillage. Besides, the water holding capacity of the soil and aeration are increased. These conditions are more favorable for soil microorganisms.
(e) *Wet tillage or puddling:* The tillage operation that is done in a land with standing water is called wet tillage or puddling. Puddling operation consists of ploughing repeatedly in standing water until the soil becomes soft and muddy (Fig. 4.8). Puddling creates an impervious layer below the surface to reduce deep percolation losses of water and to provide soft seedbed for planting rice. Puddling is done in both the directions for the incorporation of green manures and weeds.

Fig. 4.8 Wet tillage or puddling

Wet tillage destroys the soil structure and the soil particles that are separated during puddling settle later. Wet tillage is the only means of land preparation for transplanting semiaquatic crop plant such as rice. Planking after wet tillage makes the soil level and compact. Puddling hastens transplanting operation as well as establishment of seedlings. Wet land ploughs or worn out dry land ploughs are normally used for wet tillage.

4.11.2 Depth of Ploughing

The desirable depth of ploughing is 12 to 20 cm for field crops. The ploughing depth varies with effective root zone of the crop. The depth of ploughing is 10–20 cm for shallow-rooted crops and 15–30 cm for deep-rooted crops.

4.11.3 Number of Ploughing

The number of ploughing depends on soil conditions, time available for cultivation between two crops, and type of cropping systems. Zero tillage is practiced in rice fallow pulses. Minimum number of ploughing is taken up at optimum moisture level to bring favorable tilth depending on need of the crop.

4.11.4 Time of Ploughing

The optimum soil moisture content for tillage is 60% of field capacity.

4.12 Modern Concepts in Tillage

Conventional tillage involves primary tillage to break open and turn the soil followed by secondary tillage to obtain seedbed for sowing or planting. With the introduction of herbicides in intensive farming systems, the concept of tillage has been changed. Continuous use of heavy ploughs creates hardpan in the subsoil and results in poor infiltration. It is more susceptible to runoff and erosion. It is capital intensive and increases soil degradation. To avoid these ill effects, modern concepts on tillage is in rule.

4.12.1 Minimum tillage

It aims at reducing tillage operations to the minimum necessity for ensuring a good seedbed. The advantages of minimum tillage over conventional tillage are:

- The cost and time for field preparation are reduced by reducing the number of field operations.
- Soil compaction is comparatively less.
- Soil structure is not destroyed.
- Water loss through runoff and erosion is minimum.
- Water storage in the plough layer is increased.

Tillage can be reduced in two ways:

- By omitting operations which do not give much benefit when compared to the cost
- By combining agricultural operations like seeding and fertilizer application

The minimum tillage systems can be grouped into the following categories:

(i) *Row zone tillage*
Primary tillage is done with moldboard plough in the entire area of the field; secondary tillage operations like discing and harrowing are reduced and done only in row zone.

(ii) *Plough plant tillage*
After the primary tillage, a special planter is used for sowing. In one run over the field, the row zone is pulverized, and seeds are sown by the planter.

(iii) *Wheel track tillage*
Primary ploughing is done as usual. Tractor is used for sowing; the wheels of the tractor pulverize the row zone in which planting is done.

In all these systems, primary tillage is as usual. However, secondary tillage is replaced by direct sowing in which sown seed is covered in the row zone with the equipment used for sowing.

4.12.2 Zero tillage (No Tillage)

In this, new crop is planted in the residues of the previous crop without any prior soil tillage or seedbed preparation, and it is possible when all the weeds are controlled by the use of herbicides. Zero tillage is applicable for soils with a coarse-textured surface horizon, good internal drainage, high biological activity of soil fauna, favorable initial soil structure, and an adequate quantity of crop residue as mulch. These conditions are generally found in Alfisols, Oxisols, and Ultisols in the humid and subhumid tropics.

Till Planting

Till planting is one method of practicing zero tillage. A wide sweep and trash bar clears a strip over the previous crop row, and planter opens a narrow strip into which seeds are planted and covered. Here, herbicide functions are extended. Before sowing, the vegetation present has to be destroyed for which broad-spectrum nonselective herbicides like glyphosate, paraquat, and diquat are used.

Advantages

- Zero-tilled soils are homogenous in structure with more number of earthworms.
- Organic matter content increases due to less mineralization.
- Surface runoff is reduced due to presence of mulch.

Disadvantages

- Higher amount of nitrogen has to be applied for mineralization of organic matter in zero tillage.
- Perennial weeds may be a problem.
- High number of volunteer plants and buildup of pests.

4.12.3 Stubble Mulch Tillage or Stubble Mulch Farming

Soil is protected at all times either by growing a crop or by leaving the crop residues on the surface during fallow periods (Fig. 4.9). Sweeps or blades are generally used to cut the soil up to 12–15 cm depth in the first operation after harvest, and depth of cut is reduced during subsequent operations. When large amount of residue is present, a disc-type implement is used for the first operation to incorporate some of the residues into the soil. This hastens the decomposition but still keeps enough residues on top soil.

Two methods for sowing crops in stubble mulch tillage are:

1. Similar to zero tillage, a wide sweep and trash bars are used to clear a strip, and a narrow planter shoe opens a narrow furrow into which seeds are placed.
2. A narrow chisel of 5–10 cm width is worked through the soil at a depth of 15–30 cm leaving all plant residues on the surface. The chisel shatters the tillage pans and surface crusts. Planting is done with special planters.

Disadvantages of Stubble Mulch Farming

- The residues left on the surface interfere with seedbed preparation and sowing operations.
- The traditional tillage and sowing implements or equipments are not suitable under these conditions.

Fig. 4.9 Stubble mulch farming

4.12.4 Conservation Tillage

The major objective is to conserve soil and soil moisture. It is a system of tillage in which organic residues are not inverted into the soil such that they remain on surface as protective cover against erosion and evaporation losses of soil moisture. If stubble forms the protective cover on the surface, it is usually referred to as stubble mulch tillage. The residues left on soil surface interfere with seedbed preparation and sowing operations. It is a year-round system of managing plant residue with implements that undercut residues, loss the soil, and kill the weeds.

Advantages

- Energy conservation through reduced-tillage operations.
- Improve the soil physical properties.
- Reduce the water runoff from fields.

4.13 Precision Tillage

Precision tillage was described by Carter and Tavernetti (1968) wherein tillage depth was precisely specified to reach and disturb a compacted soil (Fig. 4.10). In order to apply site-specific deep tillage, cost-effective method is needed to minimize cost associated with identifying compacted zones in field.

Fig. 4.10 Precision tillage implement having additional feature of precision fertilizer placement

As the concept of GPS-based precision agriculture has gained acceptance, the idea of precision tillage has evolved to include real-time control of a "smart" tillage tool (Scarlett et al. 1997) and variable-depth deep tillage (Raper 1999). Precision deep tillage is attractive from the standpoint of eliminating unnecessary tillage. The loosening effect of subsoiling has been found to be temporary, suggesting that regular deep tillage would be required to achieve beneficial results.

Synthetic aperture radar (SAR) provides a remote sensing tool to measure soil moisture that is not compromised by cloud cover. LANDSAT satellite images are one common source for multispectral remote-sensed data from farm fields. Unfortunately, image availability is severely limited by time interval between satellite passes and the occurrence of cloud cover (Wells et al. 2000). Multispectral imagery can be used for determining soil surface conditions, including moisture content (Everitt et al. 1989). Even though remote sensing methods are generally imprecise in quantifying absolute soil moisture content, they may be useful in identifying areas of poor drainage that may indicate compacted soil. When this information is combined with crop yield maps and known field traffic patterns, areas of suspected soil compaction can be identified and, if necessary, verified by a cone penetrometer before applying precision deep tillage.

Advantages of Precision Tillage

- Maximizes existing soil moisture and increases water infiltration to create a warm, consistent seedbed with uniform seed-to-soil contact.
- Allows smooth planter operation for better performance.
- Warm, consistent seedbed for quick start and robust germination.
- Increases water infiltration and maximizes existing soil moisture.
- Some precision tillage machinery works with dry, liquid, or NH_3 fertilizers, and their row-depth control allows a precise placement of multiple nutrients at adjustable depths for efficient uptake timing.
- Independent row-depth control for precise placement and optimal uptake timing.
- Precisely places multiple nutrients at variable depths.
- Create an ideal environment throughout the growing cycle by shattering compaction in the root zone, eliminating subsoil voids, and creating twice as many beneficial pores for improved belowground development.
- Shatters root zone compaction, eliminates subsoil voids, and creates beneficial pores.
- Precision-tilled zone creates warming effect and an ideal environment throughout the season.
- Early root development during vital stages and late-stage development to improve yield results.

4.14 Conclusions

Precision soil sampling helps farmers target nutrient application. Precision agriculture allows modern producers to manage within fields rather than managing the whole field. Nutrients in any field vary due to topography, soil properties, and past management (manure application patterns, crop history, etc.). To account for this variability, farmers will need more than one soil sample in each field. A more intensive sampling scheme must be performed, through either grid or zone sampling. Looking to the future, on-the-go sensors, whether attached to tractors or unmanned aerial vehicles, may also increase accuracy and decrease the cost of soil sampling.

The utilization of large heavy equipment in many places has resulted in excessive compaction of soil that has been associated with decreased crop yield. It is relatively common for farmers who face this kind of problem to subsoil fields where compaction is suspected and/or heavy vehicles have operated. Technologies like soil sensors and remote sensing, mapping, and decision support systems in relation to soil physical and mechanical properties can be applied to achieve precision in agricultural soil tillage. Precision in tillage will lead to optimum application of input resources into crop production as well as optimization of energy, time, and economic benefits.

References

Anazodo, U. G. N., Onwualu, A. P., & Watts, K. C. (1991). Evaluation of alternative tillage systems in the absence of herbicides for maize production in a savannah loamy sand. *Journal of Agricultural Engineering Research, 49*, 259–272.

Carter, L. M., & Tavernetti, J. R. (1968). Influence of precision tillage and soil compaction on cotton yields. *Transactions of the ASAE, 11*(1), 65–0067.

Clay, D. E., Carlson, C. G., Brix-Davis, K., Oolman, J., & Berg, R. (1997). Soil sampling strategies for estimating residual nitrogen. *Journal of Production Agriculture, 10*, 446–452.

Clay, D. E., Kitchen, N., Carlson, C. G., Kleinjan, J. L., & Tjentland, W. A. (2002). Collecting representative soil samples for N and P fertilizer recommendations. *Crop Management, 1*(1), 0–0.

Everitt, J. H., Escobar, D. E., Alaniz, M. A., & Davis, M. R. (1989). Using multispectral video imagery for detecting soil surface conditions. *Photogrammetric Engineering and Remote Sensing, 55*(4).

Hawkins, E., LaBarge, G., Watters, H. D., Fulton, J., & Culman, S. (2016). Developing a strategy for precision soil sampling. *C.O.R.N. Newsletter*, 37.

Raper, R. L. (1999). Subsoiling considerations for site-specific tillage. In *American Society of Agricultural Engineers meetings papers*.

Scarlett, A. J., Lowe, J. C., & Semple, D. A. (1997). Precision tillage: in-field, real-time control of seedbed quality. In *Precision agriculture'97: Papers presented at the first European conference on precision agriculture, Warwick University Conference Centre, UK, 7–10 September 1997*. Oxford; Herndon, VA: BIOS Scientific Pub., c1997.

Wells, L. G., Stombaugh, T. S., & Shearer, S. A. (2000). Application and assessment of precision deep tillage. In *2001 ASAE annual meeting* (p. 1). American Society of Agricultural and Biological Engineers.

Wollenhaupt, N. C., Wolkowski, R. P., & Clayton, M. K. (1994). Mapping soil test phosphorus and potassium for variable-rate fertilizer application. *Journal of Production Agriculture, 7*(4), 441–448.

Chapter 5
Variable Rate Technology and Variable Rate Application

Abstract In a field comprising of different or a particular species of crop, there may be noticed variability in the fertilizer need, irrigation requirement, seed rate, chemical requirement (pesticides, herbicides, weedicides), etc. However, it is a difficult task to detect the individual requirements of a crop by mere observation; therefore, precision agriculture puts forth a technology known as variable rate technology (VRT) that helps in easy and accurate detection of individual requirements of each crop thereby reducing human errors. VRT refers to a technology that enables variable rate application of materials in precision agriculture. It is the ability to adapt parameters on machine to apply, for instance, seed or fertilizer, according to the exact variation in plant growth or soil nutrients and type. The chapter looks into working of different variable rate technologies and their application in precision agriculture.

Keywords Variable rate application · Map-based VRA · Sensor-based VRA · Lime VRA · Fertilizer VRA

5.1 How Does the Technology Work?

VRA works by encompassing the variable rate control system with application equipment to apply inputs at precise location at precise time so that the site-specific application of inputs is accomplished. Different components used are the following:

- DGPS receiver
- Computer
- VRA software
- Controller

The major components of a typical map-based variable rate control system include an in-cab computer (or controller) loaded with application software, a DGPS receiver that provides vehicle position information to the computer, and an actuator that regulates material rates under direction of the computer. When the

L. Ahmad, S. S. Mahdi, *Satellite Farming*,
https://doi.org/10.1007/978-3-030-03448-1_5

equipment is operating in the held, the computer receives position information, matches the required application rate as a function of vehicle position, and then sends a set point signal to the controller that adjusts the application to the desired rate. A variable rate system may also record actual application rates along with GPS position (Sökefeld 2010). This information serves as a record of what was applied to the field and allows for review of application for future recommendation considerations.

5.2 Variable Rate Application

Variable rate application refers to the application of parameters like fertilizer use, lime application, seeding, etc. based on the area, location, soil conditions, etc.

5.2.1 Methods

Depending upon whether the VRA technology is being used with or without GPS system, there are two methods:

- Map based
- Sensor based

(a) Map-based VRA

This method of VRA makes use of a GPS receiver. It adjusts the application rate by making use of electronic map which is also known as prescription map. A prescription map is an electronic data file which has all the necessary and specific information regarding the input rates to be given in every field zone. As the applicator moves through the field (using the field position from GPS receiver), the concentration of input changes which is detected by the prescription map kept at a desired rate. Map-based VRA also uses maps based on previous measurements which are then implemented using various strategies which are based on information like soil type, soil color, topography, crop yield, remotely sensed images, etc., and this information is crop- and location-specific (Grisso et al. 2011).

(b) Sensor-based VRA

This method of VRA does not require GPS or prescription maps. The sensors installed on applicators detect the soil properties as well as crop characteristics. Thus, a continuous flow of information is collected based on the sensor report which is then transferred to the control system which calculates the input rates and sends the calculated information to the controller, which then delivers the input to the location. Sensor-based VRA has an advantage of providing the real-time data using

Table 5.1 Differences between map based- and sensor-based VRA

S. No.	Parameter	Map based	Sensor based
1	Methodology	Grid sampling – lab analyses – site-specific maps and the use of variable rate applicator	Real-time sensors – feedback control measures and the use of variable rate applicator
2	GPS/DGPS	Very much required	Not necessary
3	Laboratory analysis (plant and soil)	Required	Not required
4	Mapping	Required	May not be required
5	Time consumption	More	Less
6	Limitations	Cost of soil testing and analysis limit the usage	Lack of sufficient sensors for getting crop and soil information
7	Operation	Difficult	Easy
8	Skills	Required	Required
9	Sampling unit	2 to 3 acres	Individual spot
10	Relevance	Popular in developing countries	Popular in developed countries

the real-time sensors unlike the map-based VRA which makes use of previously collected data to vary the application rate inputs.

Major differences between map based- and sensor-based VRA (Patil and Shanwad 2009) are presented in Table 5.1.

5.3 Applications of VRA

(i) Seeding VRA
(ii) Weed control VRA
(iii) Lime VRA
(iv) Fertilizer VRA
(v) Phosphorus VRA
(vi) Nitrogen VRA

(i) Seeding VRA

Seeding VRA reduces the seed rate by identifying and sowing seeds only in those areas which fulfil the seed requirements. Sensor-based VRAs are equipped with soil organic matter (SOM) sensors which detect the different levels of organic matter content in the soil and adjust the plant population accordingly (Fig. 5.1). Also soil moisture meters are used for depth adjustments and thereby adjusting the plant population. Seed rate is varied by attaching a gear box or motor to the VRA seeder. However, planters and drills can be made into VRA seeders by adjusting the speed of seed metering drive. This will effectively change the plant population. VRA seeding is accomplished by separating or disconnecting the planter's seed meter systems from the ground drive wheel. By attaching a motor or gear box (to change speed of the ground wheel input), the seeding rate can be varied on the go.

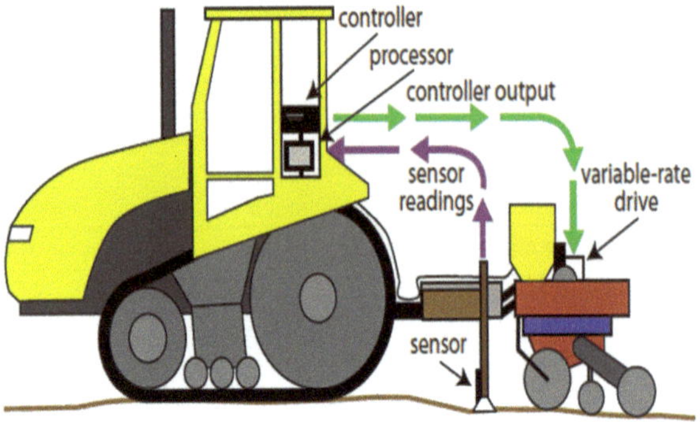

Fig. 5.1 On the go sensor measuring the soil characteristics before planting and adjusting seed rate. (Grisso et al. 2011)

Most of these devices will be matched with a prescription map and can have two or more rates. A two-rate scenario may be a system that reduces the seeding rates outside of the reach of a center pivot irrigation system, while multiple rates may be needed to adjust for soil types (water holding capacity) and organic matter.

(ii) Weed control VRA

As mentioned earlier, VRT could be either *map based* or *sensor based.* Taking into consideration the map-based weed control VRA system, a computer is required which acts as signal indicator and thus provides signal which indicates the target rate for present location. The computer used is called as *task computer*. The other requirement is a system which changes the rate of application physically, which would match the current prescribed rate.

Today's market is filled with different types of control systems adaptable to VRA. Three important categories include:

(a) Flow-based control of a tank mix
(b) Chemical-injection-based control, with the subset, chemical-injection control with carrier
(c) Modulated spraying-nozzle control system

The desire for automatically matching up the rate of applications to variations in ground speed leads to the evolution of the abovementioned systems. One of the advantages of the above systems is the elimination of error in the application that could possibly occur if there is a change in ground speed from the calibrated setup. The ability to apply variable rates (when the application rate is managed by an electronic system) would be the next logical step.

(a) Flow-based control systems

It is the simplest system of the three types mentioned above. Flow-based control system involves a combination of:

- Flow meter
- Ground speed sensor
- Control valve (servo valve)

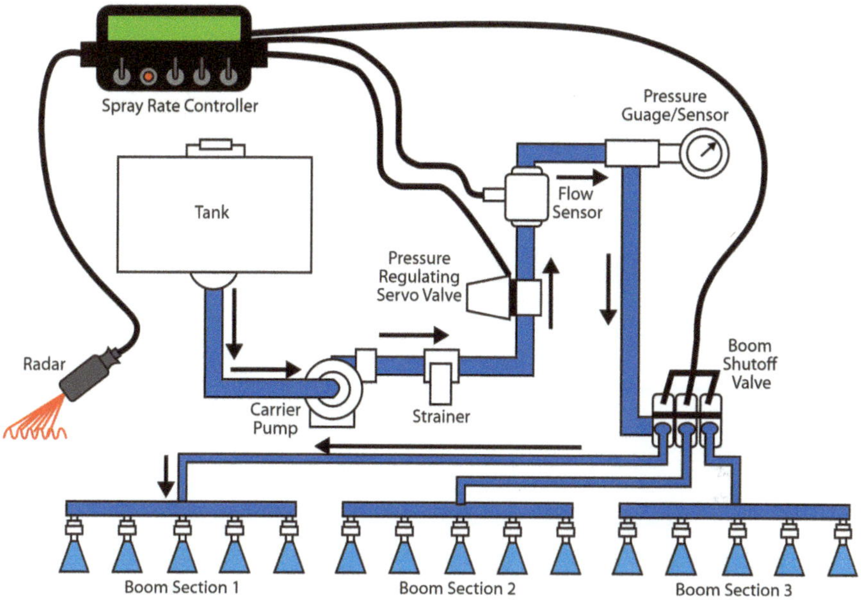

Fig. 5.2 VRA spraying system that is flow-based control system of application rate. (Grisso et al. 2011)

with an electronic controller in order to apply the desired rate of the tank mix (Fig. 5.2).

In order to calculate the appropriate flow rate for the current ground speed, there is a microprocessor in the console which uses the information regarding the sprayer width and calculates the prescribed gallons per acre. The servo valve is then opened or closed until the flow meter measurement matches the calculated flow rate. VRA can be made if there is an establishment of the communication link between this controller and a map system.

Advantages

(i) Reasonably simple and easy to use.
(ii) The system is able to make the changes in rate across the boom quickly as the control system can respond to a new rate command, which is usually very fast taking only 3 to 5 s.

Disadvantages

(i) There could be significant changes in spray droplet size and potential problems with drift as the flow sensor controls the tank mix flow by allowing variable pressure rates to be delivered to the spray nozzles.

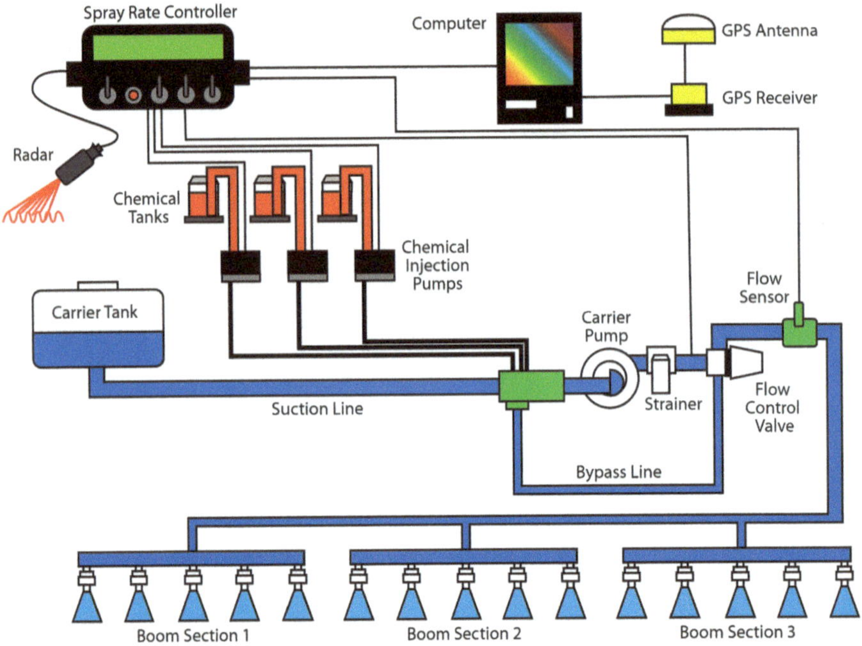

Fig. 5.3 VRA spraying system that involves chemical injection technology. (Grisso et al. 2011)

However, this is the most widely used system, the standard operation procedures of which specify that the operator must mix the chemical with the carrier in the spray tank but will generally have to deal with the leftover tank mix. Flow-based control system meets the majority of needs while giving the operators the capability of *single herbicide VRA*.

(b) Chemical direct injection systems

This system involves direct injection of chemical into a stream of water, thereby acting as an alternative approach to chemical application and control (Fig. 5.3). In order to manage the rate of chemical injection, the system makes the use of the controller and a chemical pump. The carrier constitutes the water. The rate of flow of the carrier is generally constant, and in order to accommodate for the changes in ground speed or prescribed rate, the injection rate is varied. The system can also be used for VRA, if the controller has been modified to accept the external command (from either a prescription map or a GPS signal).

Advantages

(i) Elimination of leftover tank mix
(ii) Reduction in the chances of exposure to chemicals during tank mixing
(iii) Adjustment of the constant flow of carrier (water) which provides the optimum size and distribution of spray droplets

Disadvantages

(i) Long delay between the chemical injection pump and the discharge nozzles at the ends of the boom. This means that the volume is to be applied within the hoses and attachments before the new rate has reached the nozzle. This can be a cause of delay in the rate change and would lead to "Christmas tree" application patterns as the new concentration of chemical makes its way off through the boom.

Keeping in view the above limitation, a new system worked its way out which used both carrier and injection control and came to be called as **direct chemical injection with carrier control.**

Direct Chemical Injection with Carrier Control

This system uses both carrier and injection control which change the chemical injection rate and the water carrier rate so as to respond to the speed and application rate changes. Injection pump is managed by one control loop, while the other control loop manages the servo valve which provides a matching flow of carrier (Fig. 5.4). This system delivers a mixed and constant concentration as if they have been premixed in a tank.

Advantages

(i) It shares the advantages of both the earlier mentioned systems.
(ii) No leftover mix is left.
(iii) The operator doesn't have to worry as there is no exposure to chemicals in the process of tank mixing.
(iv) Changeover from one rate to another occurs quickly.

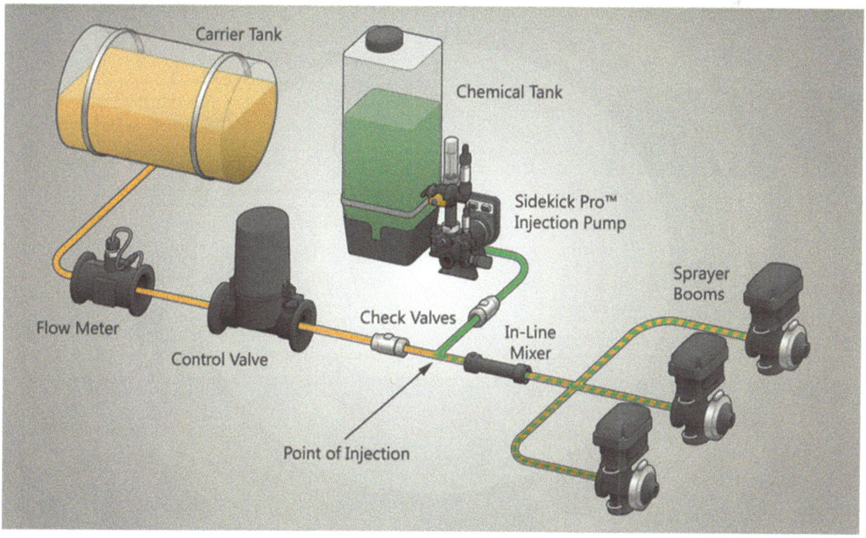

Fig. 5.4 Direct injection system

Disadvantages

(i) Complex system
(ii) Higher initial costs
(iii) Problem in delivering variable rates of liquid through the nozzle spray

(c) Modulated spraying-nozzle control systems

Modulated spraying-nozzle control (MSNC) systems with spray drift operating system are useful under a wide range of conditions. MSNC controls the duration and timing of the release through the nozzles. This system consists of the high-speed valves that control the quantity and the time for which the spray is delivered from the conventional nozzles. This system has the ability to control the rate of flow. This can further be described as:

MSNC equipped sprayers consist of sprayer nozzle assemblies that work in coordination with the direct-acting, in-line solenoid valves. The system is directed by microprocessor and an application controller that responds to signals from pressure and flow sensors (Fig. 5.5).

This system is designed to operate each nozzle at full design flow and pressure during the time for which the valve is open. The concept is however to vary the time period for which the valve stays open in order to change the rate of flow thereby changing the rate of application without changing the spray pattern. The conventional nozzle assembly is however directly mounted by a solenoid-controlled nozzle assembly.

MSNC system contains solenoids that operate at a frequency of 10 Hz. This means that in 1 s, a solenoid can ten times change its position between open and

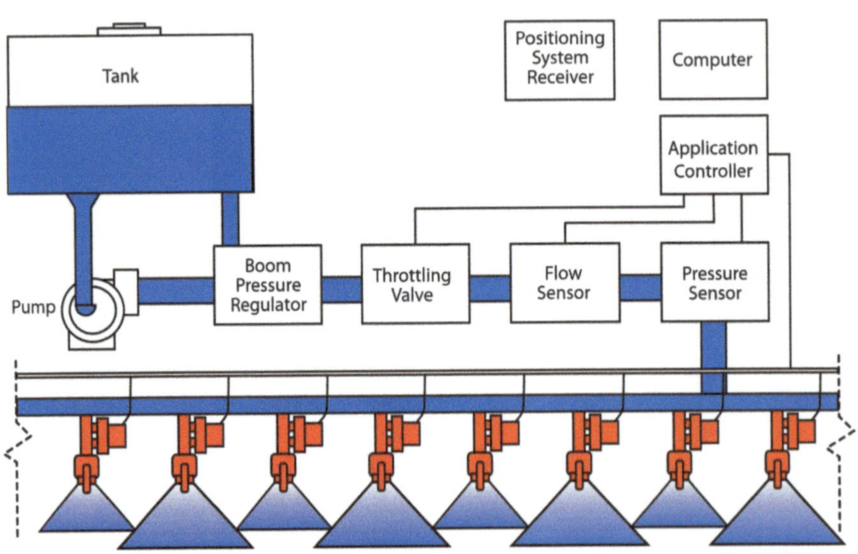

Fig. 5.5 VRA spraying system using modulated spraying nozzle control (MSNC) technology

close. This is however governed by a controller that responds to inputs from a computer processor and a set of sensors. Thus, a sequence of events (valve open, spray, valve close) takes place in one-tenth of a second.

Further, for the system to work at maximum efficiency, valve response must be quite rapid. For this, an electrical signal to each valve is used which produces one of the two flow conditions: zero flow (completely closed valve) or full flow (completely open valve). The valves operated by the solenoid take 4 milliseconds (ms) or 0.004 second (s) to respond to an electrical signal. Thus, changing the valve position from open to close and back (or vice versa) would take 8 milliseconds (ms). The amount of time for which the valve is open for flow describes the duty cycle. A minimum duty cycle can be of about 10% and a maximum of about 90%. The MSNC system can also be operated at a full open (100 percent duty cycle).

Since the rate of flow from each nozzle is governed by the time period, each flow control valve stays open; the percentage of full, rated, nozzle flow would be equal to the duty cycle expressed as percentage. The range of flow rates however varies from 9 to 1. The MSNC system can effectively be used to reduce the rates as per the need.

The MSNC system not only controls the flow rates at a given system pressure but also has the ability to increase the droplet size and reduce drift potential. This thereby manages application rates even at lowered system pressure by increasing the duty cycle of a nozzle. Opening and closing of nozzles might appear to be a risky proposition while the sprayer travels through the field. If a nozzle remains closed even for a fraction of a second, no liquid is released. This may result in depriving the certain areas of field. This is countered by using a 1/20 s (1/2 cycle) "phase shift" of adjacent nozzles. This depicts that when a nozzle is off, the adjacent nozzles need to be on. The sprayers are equipped with wide spray-angle nozzles (110 degree angle versus the more common 80 degree angle) to increase spray pattern overlap and minimize the effect of the "pulses and pauses" produced at the nozzles.

A chemical application system has the ability to control both application rate and droplet size distribution throughout the field and considers the following potential benefits:

- Produce a broader range in rates of flow with more consistent spray characteristics than the conventional sprayers.
- Alter the droplet size distribution without affecting the application rate to reduce the drift potential.
- Alter the application rates without affecting the droplet size distribution.
- Further, as the drift control is not a problem in case of fertilizer application, the MSNC systems can also be used to apply VRA nutrients. Thus, MSNC systems provide an easy way to be used as site-specific crop management methods.

– New and Developing VRA Systems

This undoubtedly is an area of rapid upcoming changes. However, the Web can easily be searched by the manufacturer's name and can effectively be used to locate product descriptions and specifications. However, each search will definitely present latest information which will guide in selecting an appropriate system for individual farming practices.

- Other Useful Devices

An automated boom section control device can beneficially be used while spraying the odd-shaped fields, grass waterways, or obstructions in a field. This system recognizes the areas of the field that should not be treated with chemicals and accordingly turn the boom (or sections of boom) on and off to prevent the chemical application over the identified areas. Also the controller can automatically turn the boom section off if the boom section is over the previously treated area thereby discarding the chances of overlap. It also manages the skips when passed back over the previously treated areas. This is because the nozzles of the implement are shut off immediately when the controller detects an overlap in order to avoid an unnecessary usage of additional chemicals. However, it requires a capital investment and may result in about 15% savings in inputs.

- Sensor-Based Devices

As the quantity of the soil organic matter influences the effectiveness of some herbicides, soil organic matter sensors can be used with VRA preplant herbicides. Such sensors do not make use of the prescription maps thus automatically adjusting the rates. In this application, the sensor is pulled or pushed through the soil by herbicide applicator.

Treating the soil uniformly can lead to the unnecessary use of the herbicides which can result in unsatisfactory weed control owing to the patchiness shown by the weeds. Remote sensing is a technique which improves weed scouting and results in better management decisions. We can easily differentiate between the weed species present in the soil based on their leaf shapes with the help of our eyes which act as remote sensors.

The remote sensing instrument collects reflectance at the field scale, and the reflectance values from the individual features are averaged over the entire pixel area within the sensor. The reflectance data from the bare ground is compared with the green weeds growing between the crops. Some sprayers are equipped to switch the device application on and off.

WeedSeeker is one example of the commercial unit having the reflectance sensor identifying the chlorophyll and a microprocessor which interprets the data, and a controller turns on the spray nozzle when a threshold level is reached. The weed seeker system is built around the close-proximity optical sensors which use the near-infrared, light reflectance to differentiate between the green vegetation, bare soil, and crop residue (Fig. 5.6). Each sensor unit consists of the optical sensor and a light source. The sensors are installed on a spray boom which is ahead of the spray nozzle and are aimed toward the ground. When the threshold level is crossed by chlorophyll reflectance signal, the controller sends the signal to the solenoid-operated valve to release the herbicide (Fig. 5.7). The system is designed such that it starts slightly before the weed is reached and turns off a bit later after crossing the

Fig. 5.6 Weed seeker, controller, and sensors

Fig. 5.7 Weed seeker-efficient cost-effective and environmentally sustainable

weed. It can operate at a speed of 3–10 mph. It reduces the chemical application significantly in the areas where the weeds are variable. The weed seeker does not distinguish between the plant types, so its agricultural use is focused on between the row applications in standing crops or on spot treatment on the fallow ground.

Boom height control is another sensor-based control device. Despite of not being a VRA device, it improves the coverage from a spray boom eliminating the streaks and improper overlaps. The sensors which are ultrasonic measure the distance to the ground (40 times per second). To make the sprayer booms follow the contours of the land automatically, responsive height adjustments are made by the control system using this information. This system shows reliable control with the average speeds more than 18 mph in all kinds of terrain.

(iii) Lime VRA

The variable rate application of lime has been regarded as the most effective and the profitable method of managing the soil pH levels. The crops show a unique response in the yield with respect to the soil pH. The yield is known to decrease with the high and the low pH levels in the soil. The farmers are forced to face the yield penalty if there is under or over application of the lime so there is a need of maintenance of accuracy which in turn gives the higher yields.

For the application of the dry chemicals, e.g., lime amendments and nutrients, e.g., nitrogen, phosphorus, and potassium, two types of applicators are used:

(a) Spinner spreaders
(b) Pneumatic applicators
(a) Spinner spreaders

The spinner spreaders which are used only vary one product at a time. The material is transferred from a conveyor belt to the hopper which passes it to the spinning discs. The application rate is managed by adjusting the gate opening and or changing the speed of conveyor

(b) Pneumatic applicators

A piped boom is used to convey the material uniformly via air stream. Centrally located bins and hoppers are installed which distribute the dry material contained in the air stream. With the help of the metering devices, single or multiple products can be blended or metered on each bin.

The methods like grid sampling and the use of on-the-go sensors can help in making a pH prescription map. On-the-go sensors work by spooning out the soil sample and forcing it against the electrode, waiting for the electrode to stabilize, and then recording the pH, eventually rinsing the whole mechanism and preparing for the next sample.

(iv) Fertilizer VRA

The VRA techniques used for the application of the fertilizer resemble the VRA techniques of liming and weed control application. The impact can be complex based on the nutrient availability, seasonal cycle, and weather condition.

(v) Phosphorus VRA

It is regarded as the second most profitable VRA activity. The grid soil tests can be used for a number of years as soil phosphorous is not clearly as transient as the soil nitrogen. Building up of soil phosphorous tests may increase the long-term economic benefits. This test is the capital investment characteristic that proves profitable to unleash the intrinsic differences in soil test phosphorous at least one point in time.

(vi) Nitrogen VRA

The adoption of VRA nitrogen (N) management by producers is low, despite the potential economic and environmental benefits of this practice. A major barrier is the recommended nitrogen fertilizer rates based on yield goal are often poorly correlated with the actual economically optimum nitrogen rates.

Nitrogen response patterns are often field and season specific and can vary widely within the same field, thus complicating the development of prescription maps. Side-by-side comparisons of uniform and VRA-N management have revealed no consistent advantages for either strategy in yields achieved, profitability, and nitrogen use efficiency by plants. In the future, a better understanding of temporal variation in nitrogen soil test levels, improved nitrogen sensing, better crop simulation models and application models, and application equipment may assist producers in capturing the benefits of VRA-N management. Real-time sensors of the crop offer the potential for VRA-N, as these systems are planned to "sense" the nitrogen needs of the crop at the time of application. These systems require well-fertilized areas in the field to get the sensor-calibrated ongoing researches to help determine if these systems would be widely employed in the future.

5.4 Conclusions

Variable rate application of cropping inputs such as seed, lime, fertilizers, and pesticides is one management strategy to address the variability that exists within agricultural fields. Adjusting the rate of inputs on the go requires sensors, electronic controllers, and mechanical drive systems, which act as the VRA technologies. Two approaches to VRA are sensor based and map based. Each method has its benefits and drawbacks. In the future, the best approach may use a combination of both.

Currently, most of the VRA technologies are commercially available, but they need an investment of time and thought of how to implement the prescription maps. The decision to use VRA and the prescriptions for varying inputs are truly site-specific. Not every farm or field will show an economic benefit from VRA, but these technologies offer opportunities for growers to increase both the production and environmental efficiencies of crop production and should be carefully evaluated.

References

Grisso, R. D., Alley, M. M., Thomason, W. E., Holshouser, D. L., & Roberson, G. T. (2011). *Precision farming tools: Variable-rate application*. London: CRC Press.

Patil, V. C., & Shanwad, U. K. (2009, December). Relevance of precision farming to Indian agriculture. In *Second national conference on agro-informatics and precision farming*.

Sökefeld, M. (2010). Variable rate technology for herbicide application herbicide application. In *Precision crop protection-the challenge and use of heterogeneity* (pp. 335–347). Heidelberg: Springer.

Chapter 6
Laser Land Leveler

Abstract Unevenness of the soil surface has a major impact on the germination, stand, and yield of crops through nutrient water interaction and salt and soil moisture distribution pattern. Land leveling is a precursor to good agronomic, soil, and crop management practices. Resource conserving technologies perform better on well-leveled and laid out fields. Farmers recognize this and therefore devote considerable attention and resources in leveling their fields properly. However, traditional methods of leveling land are not only more cumbersome and time- consuming but more expensive as well. Very often, most rice farmers level their fields under ponded water conditions. The others dry level their fields and check level by ponding water. Thus, in the process of having good leveling in fields, a considerable amount of water is wasted. This chapter gives a technical description of laser land leveler. The components of laser land leveler and the procedure for laser-guided land leveling are included in the chapter.

Keywords Laser land leveling · Topographic survey · Field plan

6.1 Land Leveling

Land leveling is used to adjust the soil surface and standardize its slope, facilitate the distribution of irrigation water, and improve field conditions for other agricultural practices (Brye et al. 2006). Land leveling for irrigation is the process of modifying the surface relief by grading and smoothing to a planned grade and to certain specifications required to facilitate or improve the uniform application of water. It is the process of flattening or modifying existing slopes or undulations rather than necessarily creating a level surface as the name may imply. The land leveling requires the earth moving over several hundred feet and should not be confused with land planning, land smoothing, or floating. These operations are usually accomplished with special equipment to eliminate minor irregularities and do not change the general topography of the land surface.

Studies have indicated that a significant (20–25%) amount of irrigation water is lost during its application at the farm fields (Cook and Peikert 1960). This problem

L. Ahmad, S. S. Mahdi, *Satellite Farming*,
https://doi.org/10.1007/978-3-030-03448-1_6

is more pronounced in the case of rice fields. Unevenness of fields leads to inefficient use of irrigation water and also delays tillage and crop establishment options. Fields that are not level have uneven crop stands, increased weed burdens, and uneven maturing of crops. All these factors tend to contribute to reduced yield and grain quality which reduce the potential farm gate income. Effective land leveling is meant to optimize water use efficiency, improve crop establishment, and reduce the irrigation time and effort required to manage crop (Rickman 2002).

6.2 Benefits of Land Leveling

Effective land leveling reduces the work in crop establishment and crop management and increases the yield and quality. Level land improves water coverage that:

- Improves crop establishment
- Reduces weed problems
- Improves uniformity of crop maturity
- Decreases the time to complete tasks
- Reduces the amount of water required for land preparation

Benefits of good leveling can be summarized as follows:

1. Basins can be formed in demanded sizes. Keeping basins small in unleveled lands will prevent land soil lose which is caused by unnecessarily made ditches.
2. Water is distributed homogeneously to the basins. Therefore, using efficiency of the present water will increase.
3. It will be easier to control water and as a result surface and deep drainages will be easier.
4. Uniform vegetation cover is provided (Meral and Temizel 2006).
5. Crop yields increase by 10–25%.
6. Water lost during the application of water is reducing by 25%.
7. Saving in irrigation water by 35–45% (Singh et al. 2008 and Chhatwal 1999).
8. Business need for irrigation is reduced by 35% (Farooq 2003).
9. This technology reduces weed problems and increases cultivable area by 3–6% (Jat et al. 2004).

6.3 Considerations in Land Leveling

There are certain prerequisites for land leveling to achieve high precision with less cost. The following should be considered for precise land leveling:

1. *Field slope*

A slope of 0–0.2% is good for optimum water flow. A small channel (e.g., 10–15 cm deep) placed in the middle of the field serves to collect the drainage water and deliver it to a drain. Good drainage characteristics add to harvest efficiency as well as create a large window for practices that need to be followed in the succeeding crop season.

2. *Infiltration*

It is important to check infiltration rates of the subsoil. In areas where a large amount of soil is moved and hardpans are removed, excessively high infiltration rate may lead to increased rates of nutrient and chemical leaching. Infiltration rate of the subsoil should at least match that of the soil surface.

3. *Cost of leveling*

The initial cost of land leveling using tractors and scarpers is high. The cost varies according to the topography, the shape of the field, and the equipment used.

4. *Maximum possible soil cut and soil fertility*

Soil cut differs from field to field with dissimilarity of subsurface fertility and soil depth. Therefore, it is necessary to encompass knowledge about immediate subsurface fertility before using laser leveling. Infertile soil should not replace fertile top layer, which may result in decreased yield in approaching seasons.

6.4 Land Leveling Techniques

Conventional Leveling

Conventionally, farmers use plankers drawn by draft animals (Fig. 6.1) or by small tractors. However, some farmers use iron scrappers or leveling boards (Fig. 6.2) connected to four-wheel tractors. Traditional land leveling includes field survey, staking and designing the field, calculation of cuts and fills, and then using a scraper and a land plan to even the land. Despite all these labor-intensive efforts, desired accuracy is not achieved. These leveling practices are crude and do not achieve a precise land leveling. Different land leveling techniques require different tools and conditions varying in operating times and accuracy (Table 6.1).

Bulldozers

Bulldozers are effective in highly undulated fields for initial leveling across larger areas, especially in cases where the terrain is rocky and the soil is hard. However, they should not be seen as presenting the perfect solution to precise leveling. Precise leveling can be achieved using advanced methods (such as application of laser technologies, as described below) that complement and refine bulldozer leveling (Fig. 6.3).

Fig. 6.1 Animal-drawn land leveler

Fig. 6.2 Tractor-drawn leveling board

Table 6.1 Time required and suitability of different land leveling techniques

Land leveling techniques	Capacity * (ha/day)	Leveling accuracy (cm)	Suitable for plot area (ha)
Animal	0.08	±4–5 cm	<0.25
Hand tractor	0.12	±4–5 cm	<0.25
Blade	0.5–1.0	±4–5 cm	<0.50
Bucket	0.5–1.0	±4–5 cm	>0.10
Laser	Up to 2 ha	±1 cm	>0.10

Jat et al. (2006)
*Compares the laser with other land levelling techniques

Fig. 6.3 Bulldozers are used for initial leveling in highly undulated fields

6.5 Laser-Controlled Land Leveling Systems

The laser-controlled system requires a laser transmitter, a laser receiver, an electrical control panel, and a twin solenoid hydraulic control valve. The laser transmitter transmits a laser beam, which is intercepted by the laser receiver mounted on the leveling bucket. The control panel mounted on the tractor interprets the signal from the receiver and opens or closes the hydraulic control valve, which will raise or lower the bucket. The components of laser-controlled land leveling systems are shown in Fig. 6.4.

Some laser transmitters have the ability to operate over graded slopes ranging from 0.01% to 15% and apply dual-controlled slope in the field.

The components of a laser-controlled land leveling systems have been described below:

(a) Four-Wheel Tractor

A four-wheel tractor is required to drag the leveling bucket. The size of the tractor can vary from 30 to 500 hp depending on the time restraints and field sizes. In Asia, tractors ranging in size from 30 to 100 hp have been successfully used with

Fig. 6.4 Components of laser-controlled land leveling systems

Fig. 6.5 Drag bucket

laser-controlled systems. It is preferable to have a four-wheel amount of soil that must be cut; it may also be necessary to plough during the leveling operation. Disc, moldboard, or tine ploughs can be used.

(b) Drag Bucket

The leveling bucket (Fig. 6.5) can be either three-point linkage mounted or pulled by the tractor's drawbar. Pull-type systems are preferred as they are easier to con-

Fig. 6.6 Electric mast

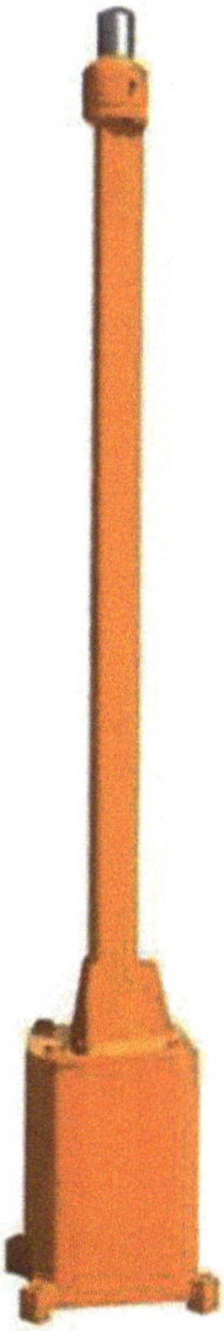

nect the tractor's hydraulic system to an external hydraulic ram than connect to the internal control system used by the three-point linkage system. Bucket dimensions and capacity will vary according to the available power source and field conditions. For example, a 60 hp tractor will pull a 2-m-wide × 1-m-deep bucket in most soil types. The design specifications for the bucket should match the available power from the tractor.

(c) Electric Mast

The electric mast (Fig. 6.6) used with the laser receiver and the control unit allows the operator to adjust remote from the cabin to the height of the laser receiver. The display of the control unit provides the operator information concerning the height of the receiver making it thus possible to make laser setups, elevation changes, and measure height differences.

(d) Laser Transmitter

The laser transmitter (Fig. 6.7) mounts on a tripod which allows the laser beam to sweep above the tractor unobstructed. With the plane of light above the field, several tractors can work from one transmitter.

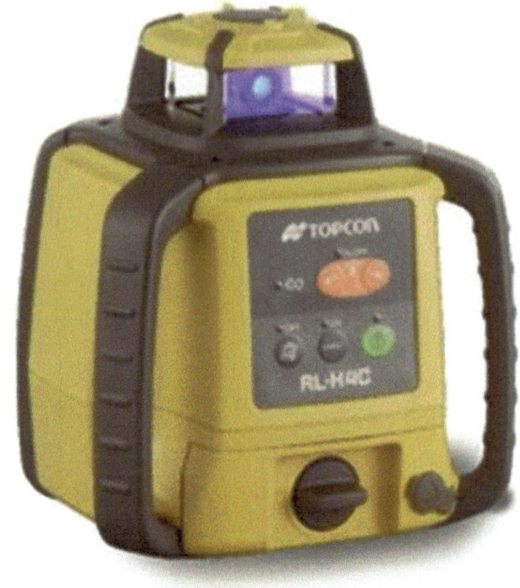

Fig. 6.7 Laser transmitter

Fig. 6.8 Laser receiver

(e) Laser Receiver

The laser receiver (Fig. 6.8) is an omnidirectional receiver that detects the position of the laser reference plane and transmits these signals to the control box. The receiver mounts on a manual or electric mast attached to the drag bucket.

Fig. 6.9 Control box

Fig. 6.10 Hydraulic control system

(f) Control Box

The control box (Fig. 6.9) accepts and processes signals from the machine mounted receiver. It displays these signals to indicate the drag bucket's position relative to the finished grade. When the control box is set to automatic, it provides electrical output for driving the hydraulic valve. The control box mounts on the tractor within easy reach of the operator. The three control box switches are On/Off, Auto/Manual, and Manual Raise/Lower (which allows the operator to manually raise or lower the drag bucket).

(g) Hydraulic Control System

The hydraulic system (Fig. 6.10) of the tractor is used to supply oil to raise and lower the leveling bucket. The oil supplied by the tractor's hydraulic pump (Fig. 6.11) is normally delivered at 2000–3000 psi pressure. As the hydraulic pump is a positive displacement pump and always pumping more oil than required, a pressure relief valve is needed in the system to return the excess oil to the tractor reservoir. If this relief valve is not large enough or malfunctions, damage can be caused to the tractor's hydraulic pump.

Fig. 6.11 Hydraulic cylinder used to supply oil to raise or lower the leveling bucket

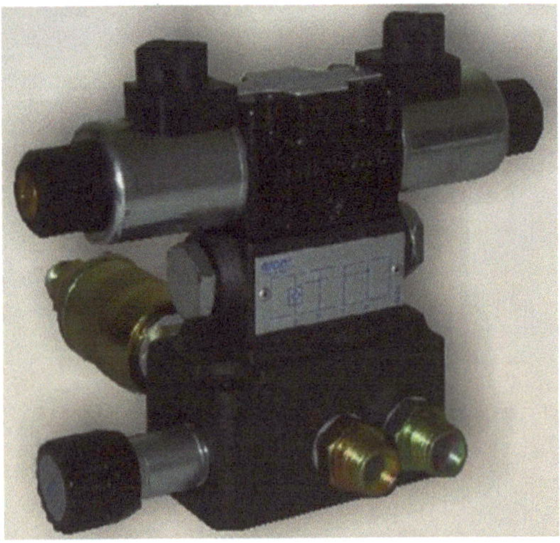

Fig. 6.12 Solenoid valve

Wherever possible, it is advisable to use the external remote hydraulic system of the tractor as this system has a built-in relief valve. Where the oil is delivered directly from the pump to the solenoid control valve, an in-line relief valve must be fitted before the control valve. The solenoid control valve, when supplied by the laser manufacturers, has a built-in relief valve.

The solenoid control valve (Fig. 6.12) controls the flow of oil to the hydraulic ram which raises and lowers the bucket. The hydraulic ram can be connected as a single- or double-acting ram. When connected as a single-acting ram, only one oil line is connected to the ram. An air breather is placed in the other connection of the ram to avoid dust contamination on the non-working side of the ram. In this configuration, the weight of the bucket is used for lowering.

The desired rate at which the bucket raises and lowers will depend on the operating speed. The faster the ground speed, the faster the bucket will need to adjust. The rate at which the bucket will raise and lower is dependent on the amount of oil supplied to the delivery line. Where a remote relief valve is used before the control valve, the pressure setting on this valve will change the raise/lower speed. Laser manufacturer supplied control valves have pressure control adjustments on both the bypass relief valve and the raise and lower valves.

When using a hydraulic ram, it should be positioned so that the ram body is connected in such a way as to push from the bucket frame rather than the depth control wheels.

6.6 Steps for Laser Land Leveling

Laser leveling requires soil to be shifted from the high points of the field to the low points in the most cost-effective way. In most situations, fields will need to be ploughed and a topographic survey undertaken before leveling commences. Various steps involved in laser land leveling are given below:

1. Ploughing the Field

Plough the field preferably from the center of the field outward. It is preferable to plough the field when the soil is moist because if the soil is ploughed dry, a significant increase in tractor power is required and large clod sizes may result. If the soil is very dry, a one-way disc or moldboard may be required. Disc harrows or tine implements are ideal for second workings. All surface residues need to be cut up or removed to aid soil flow from the bucket.

2. Conducting a Topographic Survey

Once the field is ploughed, you should conduct a topographic survey to record the high and low spots in the field. From the surveyed readings, you can then establish the mean height of the field by taking the sum of all the readings and dividing it by the number of readings taken. Then, using a field diagram and the mean height of the field, you can determine a strategy to effectively move soil from the high to low areas.

Lasers are now widely used to accomplish a topographic survey. They are very accurate, simple to use, and readily available in most countries. Recordings can be taken up to a radius of 300 meters from the transmitter. One person can operate a laser level.

The laser surveying system is made up of a laser transmitter, a tripod, a measuring rod, and a small laser receiver. A major advantage of laser surveying is the accuracy and simplicity of use, and only one person is needed.

Other Equipment

Other equipment may be needed when using a laser system for topographic surveying. This includes:

1. **Tape:** One 100 m tape. White metal tapes are more accurate, and fiberglass tapes are generally more robust.

2. **Staff:** if a measuring rod is not available. The staff is preferred as metric and upright with an E-type pattern. Check the accuracy of the scale on the staff using a steel tape because some staff faces may be out by one or more centimeters.
3. **Compass:** If direction and bearings are to be recorded, a compass will be required. The compass can be used to set magnetic north on the level and allow recordings to be taken from it.
4. **Pegs/Hammer:** Pegs, preferably painted white, are required especially for marking out a grid survey or temporary marks.
5. **Book:** A notebook is required to record all measurements and other information required to make sense of the survey work completed in the field. Enough detail should be recorded so that the levels could be rechecked if necessary. The notebook should be small enough to fit comfortably into a shirt or trouser pocket.
6. **Pencil/Eraser:** A pencil and eraser are preferable to an ink- or biro-type pen in the field. Mistakes can be easily remedied, pencils will not run out as an ink pen will, and a pencil will still function even if the paper becomes a little wet through perspiration.

Using a Laser Level

1. Open the tripod legs and adjust the individual positioning of the legs until the base plate is relatively level. Use the horizon as a visual guide to get the base plate level.
2. Attach the laser transmitter to the base plate.
3. If the laser is not self-leveling, adjust the individual screws on the base of the transmitter to get the bubble into the center of both circles. Most lasers will not rotate unless the transmitter is level.
4. Once the transmitter is level, attach the receiver to the staff and activate the sound monitor.
5. The laser is now ready to commence recording heights.

2.1. Recording Survey Measurements

All measurements should be recorded in a field book that can be easily carried in a shirt or trouser pocket. There are a number of ways to record data in a field book, but the open-field plan system is the easiest when surveying for leveling purposes.

Field Plan Format

When leveling a field or placing contour banks in a field, it is often easier and less confusing to record the data on a scaled or grid map that visually represents the field.

Field Plan

A benchmark is used so that any point in the field can be checked back against a reference if the need arises and to check the accuracy of the survey. The levels are recorded on a grid map (Fig. 6.13). Change in elevation or comparing individual readings or subtracting each reading from the mean height of the field can determine slope.

In the case of a land leveling exercise, where a cut/fill map is required, it is best to use the mean height of the field as a reference point. To determine the mean height of the field, add together all of the measured points and divide by the number of measurements.

Fig. 6.13 Benchmarks for checking the accuracy of the survey

A new map is then drawn which shows the difference between the mean height of the field and the recorded height. For example, if the mean height is 1.00 m and the actual height is 1.50 m, then there is a low point of 0.5 m. This is recorded as a positive figure, and when leveling land, this area must be filled. By doing this for the complete field, a cut/ fill map is drawn, and an estimate of the volume of soil that needs to be shifted can be calculated.

3. Leveling the Field

   Leveling a field involves the following steps:

1. The laser-controlled bucket should be positioned at a point that represents the mean height of the field.
2. The cutting blade should be set slightly above ground level (1–2 cm).
3. The tractor should then be driven in a circular direction from the high areas to the lower areas in the field (Fig. 6.14).
4. To maximize working efficiency, as soon as the bucket is nearly filled with soil, the operator should turn and drive toward the lower area. Similarly, as soon as the bucket is near empty, the tractor should be turned and driven back to the higher areas.
5. When the whole field has been covered in this circular manner, the tractor and bucket should then do a final leveling pass in long runs from the high end of the field to the lower end.
6. The field should then be resurveyed to make sure that the desired level of precision has been attained.
7. In wet areas where there is poor traction or a chance of bogging the tractor, care needs to be taken to fill the wet areas from the affected edge in a circular motion.
8. If the fields are ploughed in lands and leveling is undertaken in the areas of soil settlement in the second year, the fields should not require further major leveling works for at least 8 years.
9. Maintain or repair levees.

Fig. 6.14 Tractor should be driven in circular direction in the field from high to lower areas

6.7 Benefits of Laser Land Leveling

Laser leveling systems are commonly used in agricultural applications in Australia, Japan, and the United States. Increasingly, laser-guided systems are being used in a lesser developed country context as well. Laser leveling results in a much more level field. Accuracy can be improved by as much as 50%.

The benefits over other land leveling methods include the following:

- More level and smooth soil surface
- Reduction in time and water required to irrigate the field
- More uniform distribution of water in the field
- More uniform moisture environment for crops
- More uniform germination and growth of crops
- Reduction in seeds, fertilizer, chemicals, and fuel used in cultural operations
- Improved field traffic ability (for subsequent operations)

Yield

Research has shown a large increase in crop yield due to good field leveling. Land leveling increases yield. A large part of this increase is due to improved weed control. Improved water coverage from better land leveling reduces weeds by up to 40%. This reduction in weeds results in less time for crop weeding. A reduction from 21 to 5 labor days per hectare is achieved. This represents a reduction of up to 16 person-days per hectare decrease in the labor required for weeding.

Farm Operation
Land leveling makes possible the use of larger fields. Larger fields increase the farming area and improve operational efficiency. Increasing field sizes from 0.1 to 0.5 ha increases the farming area by between 5% and 7%. This increase in farming area gives the farmer the option to reshape the farming area that can reduce operating time by 10–15%.

Seeding Practices
Leveling reduces the time taken for planting, for transplanting, and for direct seeding. Land leveling provides greater opportunity to use direct seeding. The possible reduction in labor by changing from transplanting to direct seeding is approximately 30/person-days per hectare.

Efficiency of Water Use
Rice farmers using animals or 2-wheel tractors rely on water to accumulate in the field before starting land preparation. The average difference in height between the highest and lowest portions of rice fields in Asia is 160 mm. This means that in an unleveled field, an extra 80 mm to 100 mm of water must be stored in the field to give complete water coverage. This is nearly an extra 10% of the total water requirement to grow the crop.

Financial Benefits of Land Leveling
Although the initial cost of land leveling is an extra expense, a cash flow over a period of years shows that financial benefits do result from land leveling. The costs allow for extra ploughing and extra fertilizer in the first and second years. The benefits include reduced weeding costs of 40%.

6.8 Limitations of Land Leveling

The limitations include the following:

- High cost of the equipment/laser instrument
- Need for skilled operator to set/adjust laser settings and operate the tractor
- More efficient for regularly sized and shaped field.

6.9 Conclusions

Agricultural land leveling is of high importance in modern and intensive agriculture, especially under drought-prone conditions where water resources are limited. The technology of laser leveling is very much effective in tackling these issues. Laser land leveling is meant to optimize water use efficiency, better distribution of soil moisture, improve crop establishment, reduce the irrigation time, save irrigation water and effort required to manage crop, and reduce spatial variability.

References

Brye, K. R., Slaton, N. A., & Norman, R. J. (2006). Soil physical and biological properties as affected by land leveling in a clayey aquert. *Soil Science Society of America Journal, 70*(2), 631–642.

Chhatwal, G. R. (1999). *Encyclopaedia of Environmental Agricultural Pollution and Its Control*. New Delhi: *Anmol Publications PVT. LTD.*

Cook, R. L., & Peikert, F. W. (1960). A comparison of tillage implements. *The Journal of American Society of Agricultural Engineers, 31*, 221–214.

Farooq, N. (2003). *Conservation Agriculture and The rural Women Breaking Barriers For a new Horizon*. District Sheikhupura: Chairperson, Community Welfare Society.

Jat, M. L., Chandna, P., Gupta, R., Sharma, S. K., & Gill, M. A. (2006). Laser land leveling: A precursor technology for resource conservation. In *Rice-Wheat consortium technical bulletin series* (Vol. 7, p. 48).

Jat, M. L., Pal, S.S., Subba Rao, A.V. M., Sirohi, K., Sharma, S. K., & Grupta, R. K. (2004). Laser land levelling: The precursor technology for resource conservation in irrigated ecosystem of India. Abstracts, National Conference on conservation agriculture: Conserving resources and enhancing productivity.

Meral, R., & Temizel, K. (2006). Irrigation applications and efficient water use in Rice production. Kahramanmaras Sutcu imam univ. *Journal of Sciene and Engineering, 9*(2), 104–109.

Rickman, J.F. (2002). *Manual for laser land leveling* (Vol. 5). Rice-Wheat Consortium for the Indo-Gangetic Plains.

Singh, Y., V. P. Singh, B. Chauhan, A. Orr, A. M. Mortiner, D. E. Johnson, & B. Hardy (2008). Direct seeding of rice and weed management in the irrigated rice-wheat cropping system of the Indo-Gangetic Plains. los Banos (Philippines). *International Rice Research Institute, and Pantnagar (India):* Directorate of Experiment Station, G. B. Pant University of Agriculture and Technology (p. 272).

Chapter 7
Site-Specific Nutrient Management

Abstract Site-specific nutrient management is a component of precision agriculture and can be used for any field or crop. It combines plant nutrient requirements at each growth stage and the soil's ability to supply those nutrients and applies that information to areas within a field that require different management from the field average. Site-specific management allows for fine-tuning crop management systems along with 4R Nutrient Stewardship – the right source, rate, time, and place of nutrient use. Different concept and approaches of nutrient management of SSNM have been elaborated in this chapter. The chapter includes the present approach and constraints in the adoption of SSNM particularly in India.

Keywords Site-specific nutrient management · Nutrient management · SSNM tools · Chemigation and fertigation

7.1 Introduction

Understanding the principles of soil fertility is vital to efficient nutrient management, crop production, as well as environmental protection. There are 17 chemical elements known to be essential for plant growth, and 14 of these elements come from the soil (Table 7.1). Each essential plant nutrient is needed in different amounts by the plant, varies in mobility within the plant, and varies in concentration in harvested crop components. It is useful to know the relative amount of each nutrient that is needed by a crop and the relationship to amounts removed with crop harvest.

To be classified as essential, the element needs to meet the following criteria:

1. The plant cannot complete its life cycle (seed to new seed) without it.
2. The element's function cannot be replaced by another element.
3. The element is directly involved in the plant's growth and reproduction.

L. Ahmad, S. S. Mahdi, *Satellite Farming*,
https://doi.org/10.1007/978-3-030-03448-1_7

Table 7.1 Essential plant elements, source, roles, and relative quantities in plant

Element	Source	Role in Plant	Concentration
Carbon (C)	Air	Constituent of carbohydrates; necessary for photosynthesis	45%
Oxygen (O)	Air/water	Constituent of carbohydrates; necessary for respiration	45%
Hydrogen (H)	Water	Maintains osmotic balance; important in many biochemical reactions; constituent of carbohydrates	6%
Nitrogen (N)	Air/soil	Constituent of amino acids, proteins, chlorophyll, and nucleic acids	1–5%
Potassium (K)	Soil	Involved with photosynthesis, carbohydrates translocation, and protein synthesis	0.5–1%
Phosphorous (P)	Soil	Constituent of proteins, coenzymes, nucleic acids, and metabolic substrates; important in energy transfer	0.1–0.5%
Magnesium (Mg)	Soil	Enzyme activator; component of chlorophyll	0.1–0.4%
Sulfur (S)	Soil	Component of certain amino acids and plant proteins	0.1–0.4%
Chlorine (Cl)	Soil	Involved with oxygen production and photosynthesis	0.01–0.1%
Iron (Fe)	Soil	Involved with chlorophyll synthesis and in enzyme electron transfer	50–250 ppm
Manganese (Mn)	Soil	Controls several oxidation-reduction systems and photosynthesis	20–200 ppm
Boron (B)	Soil	Important in sugar translocation and carbohydrate metabolism	6–60 ppm
Zinc (Zn)	Soil	Involved with enzymes that regulate various metabolic activities	25–150 ppm
Copper (Cu)	Soil	Catalyst for respiration; component of various enzymes	5–20 ppm
Molybdenum (Mo)	Soil	Involved with nitrogen fixation and transforming nitrate to ammonium	0.05–0.2 ppm
Nickel (Ni)	Soil	Necessary for proper functioning of urease and seed germination	0.1–1 ppm

Pagani et al. (2013)

7.2 Nutrient Management

Nutrient management is a major component of soil and crop management system. Knowing the required nutrients for all stages of growth and understanding the soil's ability to supply those needed nutrients are critical to profitable crop production. The importance of fertilizer nutrient application to crops for achieving higher yield and to sustain the soil fertility on a long-term basis is well known. Nutrient management implies managing all nutrient sources, fertilizers, organic manures, waste materials suitable for recycling nutrients, soil reserve, and biofertilizers in such a way that yield is not knowingly jeopardized, while every effort is made to minimize losses of nutrients to the environment. The nutrient management is a challenge to maintain and increase sustainable crop productivity which could meet the demands for food and raw materials, and to enhance the quality of land and water resources.

Improving nutrient management strategies is equally important as breeding for high yielding. Several approaches have been used for fertilizer nutrient management in the past such as recommending a fixed rate based on nutrient response curve or recommendations based on balance sheet method. Fertilizer nutrient management based on soil test value, i.e., soil testing, is one of the most widely used methods in India, where fertilizer recommendations are made based on initial soil test value. Among the other approaches, the target yield approach has got popularity in India. This method not only estimates soil test-based fertilizer dose but also the level of yield the farmer can achieve with a particular dose. The success of any nutrient management strategy depends on the spatial scale for which decisions are made; hence, nutrient management concept can distinctly be defined for two broad domains:

1. Package approach based on blanket recommendation over wider area
2. Site-specific nutrient management

Conventionally, blanket fertilizer recommendations are often applied over larger areas without taking into account the wide variability in soil nutrient supplies, fertilizer efficiency, and site-specific crop nutrient requirements within each recommendation domain. Package approach based on blanket recommendation including fixed rate nutrient application, soil test-based approaches, and balanced sheet method assumes that crop needs for nutrients are constant among years and over large areas. But crop growth and demand for nutrient are strongly influenced by climate and crop-growing conditions which can vary greatly among locations, seasons, and years. Hence, current nutrient management strategies suffer from a number of limitations:

- They do not assist farmers in dynamic nutrient management.
- Do not consider need of crop grown in different seasons and places.
- Lead to very low nutrient use efficiency, e.g., N-0.3-0.5; P-0.15-0.2, and K-0.80.
- Give poor yields because of low nutrient use efficiency.
- Poor returns.
- Soil and water pollution due to excess nutrient losses.

To eliminate wastage of fertilizer and increase farmer's income, the need was felt to make adjustments in fertilizer amount and timing with respect to the location and season-specific needs of crop. It is argued that a sustainable nutrient management strategy must ensure high and stable overall productivity with optimum economic returns and sufficient nutrient supply for potential yield increases with minimum wastage of nutrients. This can be achieved if exogenous nutrient supply is matched with nutrient supplies from soil and crop demands. Therefore, a site- specific nutrient management (SSNM) strategy is required to increase efficiency of nutrients and crop productivity (Janssen et al. 1990; Cassman et al. 1996). Nutrient management strategies based on more site-specific approaches can offer opportunities for efficient utilization of nutrients applied through fertilizers or organic sources.

7.3 Site-Specific Nutrient Management

Site-specific management can be thought of as a series of layers of information about each field, as depicted in Fig. 7.1. Each time a measurement is made (soil tests, scouting reports, yield data, etc.), another layer of information is added. Over time, multiple layers of information are added and become part of the database that can guide future crop management decisions. By geo-referencing each data point to its precise geographic location, these data layers can be "stacked" for analysis to determine the relationship between layers for any point in the field. For example, the relationship between nitrogen rate applied and yield obtained might be determined, and then its variability mapped as an additional "calculated" layer of information.

Site-specific nutrient management is an approach of feeding crops with nutrients when needed. The application and management of nutrients are dynamically adjusted to crop needs of the location and season. It is a low technology plant-based approach for managing the nutrient requirement of crops. It provides principles and tools for supplying nutrients to crops when needed to achieve high yields while optimizing the use of nutrients from indigenous sources. This concept can be applied to any field and any crop. While most often, in relation to the use of computer and satellite technology, the site-specific nutrient management does not require special equipment nor does it require a large farming operation, the technology tools certainly expand the capabilities for using site-specific nutrient management.

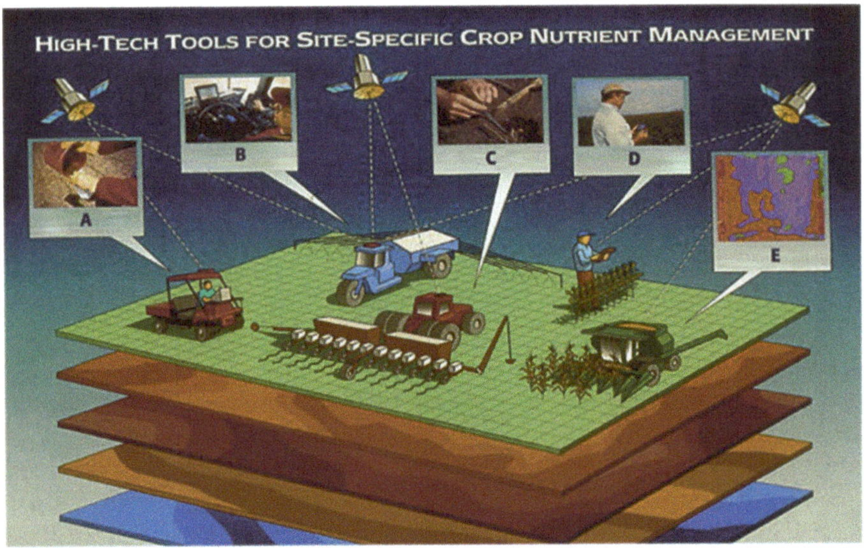

Fig. 7.1 Layers of information in SSNM. (A). Grid sampling, guided by GPS, provides more accurate soil test data. (B). Variable rate fertilizer application can improve efficiency. (C). Variable rate seeding, variety changes, and starter fertilizer can adjust for soil properties and productivity. (D). Crop scouting with new digital technologies improves field records. (E). On-the-go yield monitors can quickly track variability in the field

Site-specific crop and soil management are really a repackaging of management concepts that have been promoted for many years. It is basically taking a systematic approach to applying sound agronomic management to small areas of a field that can be identified as needing special treatment. The components of site-specific management may not be new, but we have the capability with new technology to use them more effectively. Site-specific management includes practices that have been previously associated with maximum economic yield (MEY) management, best management practices (BMPs), as well as general agronomic principles. The systematic implementation of these practices into site-specific systems is probably our best opportunity to develop a truly sustainable agriculture system.

7.4 Concept and Approaches of Nutrient Management

Site-specific nutrient management is a widely used term all over the world with reference to addressing nutrient differences which exist within fields and making adjustments in nutrient application to match those locations or soil differences. The earliest use of the term "site-specific" comes from the late 1920s in the United States, when scientists at the University of Illinois were providing recommendations on the application of lime to acidic soils. In the mid- to late 1980s, the development of GIS and GPS technology and the associated use of this technology in agriculture to map field variability fostered the use of "site-specific management" approaches in the developed world. The International Plant Nutrition Institute (IPNI) started a site-specific soil testing and assessment program in India, which can be best described as a systematic approach to soil fertility evaluation. It outlined a local approach for evaluating soil fertility and making fertilizer recommendations which could be implemented by any institute or state research group (Portch and Hunter 2002). This method involves two major steps in the laboratory, prior to any field research being carried out.

The first is to conduct a series of fixation reaction studies on nutrient elements with soil, determining the ration of nutrient supply versus nutrient applied with a specific soil. This information helps to indicate how much nutrient is required to be added to a soil to bring it to a level which is more than adequate for maximum growth but less than that which could be toxic or out of balance with other plant nutrients.

The second step involves a greenhouse study where sorghum is grown as an indicator plant with the complete nutrient treatment, including all nutrients required for growth provided in the fertigation system and deletion treatments for each of the individual nutrients. The result is biomass sample data which will provide an indication of when a potential positive or negative response is to be expected. This information is then used to proceed to field trials with the specific crop or crop's providing the necessary insight as to what nutrients need to be considered in the evaluation. The absence of any response to a nutrient in the greenhouse trial helps to eliminate it from further assessment in the field trials, bringing the focus on the deficient nutrients. Using yield goals and the soil analysis, the lab is in a position to provide fertilizer recommendations covering the full range of nutrients required, including macro, secondary, and micronutrients (Portch and Stauffer 2005). This procedure was introduced in both India and China, with a national soil fertility lab, using the methods now operating in Beijing.

7.5 Present Approach for SSNM

Site-specific nutrient management (SSNM) is a plant-based approach for managing the nutrient needs of crops. It provides principles and tools for "feeding" plants with nutrients when needed to achieve high yields while optimizing the use of nutrients from indigenous sources. The present approach of SSNM is based on the following steps:

1. Establish an attainable yield target for farmers' fields

   The yield target in the SSNM approach is directly used to calculate fertilizer rates for farmer's field. It must be reasonably attainable by farmers. A yield target higher than one attained by farmers would lead to recommendations of more fertilizer than required for high use efficiency and profit. A yield target below that realistically attained by farmers could result in suboptimal yield and profit. Grain yield from fully fertilized plot with no nutrient limitations and good management (like NPK plot or NPK plus micronutrient plot in the nutrient omission plot technique) can be used to estimate the yield target.

2. Approximate a fertilizer N rate and formulate dynamic N management

   The difference between the yield target and N-limited yield (i.e., yield with no N fertilizer and no limitation of other nutrients) provides an estimate of anticipated crop response to fertilizer N. The estimated yield response to fertilizer N and a targeted efficiency for fertilizer N use are used to approximate the total requirement of crop for fertilizer N which is dynamically apportioned among multiple times of application to best match the crop's need for N.

3. Estimate field-specific nutrient-limited yields

   Nutrient-limited yields were determined by the nutrient omission plot technique. The K-limited yield is determined in a K omission plot receiving no K fertilizer but sufficient N and P to ensure they do not limit yield. The P-limited yield is determined in a plot receiving no P fertilizer but sufficient supply of other nutrients.

4. Regulation of P and K rate

   Crops need for fertilizer P based on a comparison of yield target and P-limited yield, whereas the crops need for fertilizer K is based on a comparison of the yield target and K-limited yield. The SSNM approach advocates sufficient use of fertilizer P and K to both overcome P and K deficiencies and maintain soil P and K fertility.

7.6 Delivering the SSNM Approach to Farmers in India

A major aspect of any research program in plant nutrition is determining how this information will ultimately be used to positively impact the farmer's practices. India has a vast population of farmers, many of which are poorly educated and have very limited ability to interpret new technologies. In fact, an NSS survey reported in the Hindu Business Line on 2 August 2005 reported that most farmers get their information from their progressive neighbors (16.7%), input dealers (13.1%), or from the radio (13.0%). Government extension workers were noted by only 5.7% of farmers as a source for accessing information on modern technology. This dependence on progressive neighbors is not uncommon and indicates that many farmers are making business decisions based on the experiences and judgment of their neighbors. Radio programs provide much more general information, requiring a certain degree of interpretation relative to an individual farm. However, the role that input dealers play indicates that these farmers are seeking out a person who has contact with their location in the region and understands the soils, crops grown, yield potentials, and the emerging issues. Another area in which IPNI in collaboration with IRRI and fertilizer industry is making good strides is to consolidate the complex and knowledge-intensive SSNM information into a simple delivery system enabling farmers to rapidly implement SSNM. Such delivery system developed in 2008 by IRRI and called "nutrient manager" is an easy-to-use, interactive computer-based decision tool (Buresh 2008). The tool endeavors to acquire necessary information required for decision- making on nutrient management through a series of easy-to-answer questions which essentially mask the rigor of SSNM principles from end users while maintaining robustness of the process. These decision tools consist of about ten multiple-choice questions that can easily be answered by an extension worker or farmer. Based on the response to the questions, a fertilizer guide line with amounts of fertilizer by crop growth rate is provided for the rice field. They accommodate transplanted and direct seeded rice, inbred or hybrid varieties with a range of growth durations, as well as rice-growing seasons to provide a range of options to wide-spectrum of farmers.

Recently, IPNI has successfully tried geographical information system (GIS) mapping approach to measure spatial variability in nutrient status and used such maps as a site-specific fertilizer recommendation tool to positively impact rice yields in farmer's fields. This mapping is based on two factors: (a) nutrient content of cultivated soils varies spatially due to variation in genesis, topography, cropping history, fertilization history, and resource availability and (b) soil testing of all holdings to estimate native fertility levels to ensure appropriate recommendation is a logical step, but we do not have adequate infrastructure to accomplish this task. The process of soil fertility mapping involved geo-referenced soil sampling and using the soil analysis data in a GIS platform to develop surface maps of analyzed soil parameters across the study area. The spatial variability maps created by combining the location information of sampling points and analyzed soil parameter are capable of predicting soil parameter values of un-sampled points. This is possible because the interpolation technique used in the GIS platform creates smooth surface map of the study area utilizing point information (geographic location and corresponding soil parameters), where each point on the map has soil parameter value associated with it. The possibility of using such maps as fertilizer decision support tool to guide nutrient application in site-specific mode is being explored in several studies under IPNI search initiative. Besides delineating the fertility management zones, these maps can give a clear visual indication within study area; these maps can give a clear visual indication of changing fertility scenario at village level with time, which is important for nutrient management planning. Besides the logistical and economic advantages of implementing such a system, once established, the technique can create an effective extension tool where field agents work more directly with farmers. Thus, farmers become more aware of how their fields rank within the landscape in terms of basic soil fertility which in turn enables a system of more rational use of fertilizer application.

7.7 SSNM Tools

(i) *Optical sensors*

Farmers and extension agents can use optical sensors (to develop SSNM recommendations, particularly for N. Optical sensors measure reflectance from the leaves to generate a vegetative index called NDVI (normalized difference vegetation index), which measures the nutrient status of the plants based on their size and color (green versus yellow). The original technology was developed for large farms; however, a small handheld version that costs a fraction of the original technology (approximately USD 500) is now commercially available (Crain et al. 2012).

The application of optical sensor-based nutrient management requires a local calibration of the sensor for a given nutrient, crop, and region. This calibration relates the grain yield of the crop to the NDVI readings. Once calibration is complete, optical sensors require:

1. Establishment of a reference strip in the farmer's field that will receive a non-limiting amount of N.
2. Collection of an NDVI reading in the reference strip and in the field area where the farmer needs to know how much N should to be applied.
3. The NDVI readings collected from these two areas in the field together with the date of planting and date of sensing are entered in a mathematical model developed for each region.

(ii) *Leaf color chart*

LCC, jointly developed by the IRRI and Philippine Rice Research Institute, is an innovative cost-effective tool for real-time or crop-based nitrogen management in rice, wheat, and maize. It is a visual and subjective indication of plant nitrogen deficiency and is an inexpensive, easy-to-use, and simple alternative to chlorophyll meter and SPAD.

(iii) *Software for SSNM*

Computer- or mobile phone-based tools are increasingly used to facilitate improved nutrient management practices in farmers' fields, especially in geographies where blanket fertilizer recommendations prevail. These tools provide small-scale maize, rice, and wheat farmers with crop and nutrient management advice customized to their farming conditions and needs.

(iv) *Soil plant analytical device (SPAD)*

It is a device which is used to provide a rapid, accurate, cost-effective, and understandable analytical test results of soil and leaf samples. Soil and leaf testing in turn provides the critical information required to ensure all crop nutrition decisions are accurate, efficient, cost-effective, and responsible.

7.8 Chemigation and Fertigation

Chemigation is the injection of any chemical such as nitrogen, phosphorus, or pesticide into irrigation water and applied to the land using the irrigation system (Fig. 7.2). The proper use of chemigation is recognized as a best management practice (BMP) for irrigated agriculture. However, in order to use chemigation to benefit your crop production, you have to understand the equipment and management requirements and limitations.

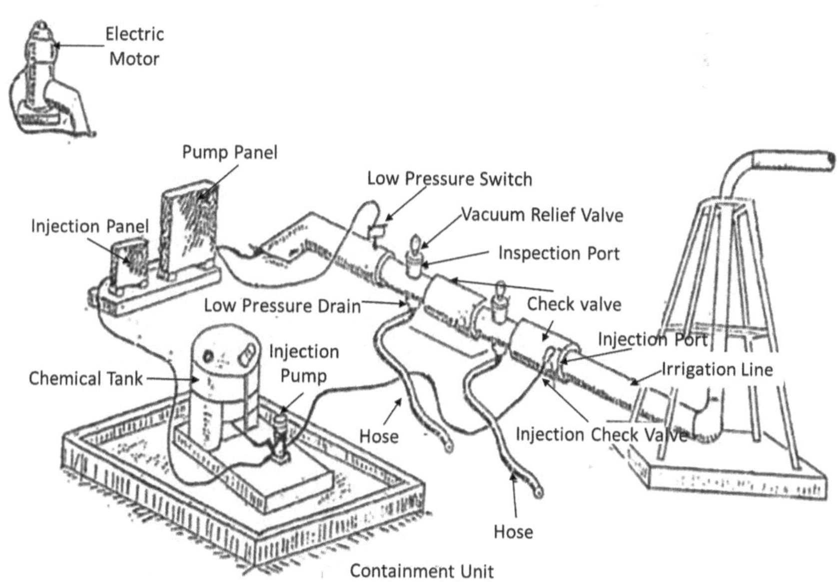

Fig. 7.2 Chemigation equipment arrangement when applying a pesticide with an irrigation system connected to an irrigation well

Advantages of Chemigation

1. *Uniformity of application.* With a properly designed irrigation system, both water and chemicals can be applied uniformly, resulting in excellent distribution of the water-chemical mixture.
2. *Precise application.* Chemicals can be applied where they are needed and in the correct concentrations.
3. *Economics.* Chemigation is usually less expensive than other application methods and often requires smaller amounts of chemicals.
4. *Timeliness.* Chemigation can be carried out when other methods of application might be prevented by wet soil, excessive wind, lack of equipment, and other factors.

The safety of the operator is not in the field continuously during applications, so there is less human contact with chemical drift and less exposure during frequent tank fillings and other tasks.

Disadvantages of Chemigation

1. Skill and knowledge required. Chemicals must always be applied correctly and safely. Chemigation requires skill in calibration, knowledge of the irrigation and chemigation equipment, and an understanding of the chemical and irrigation scheduling concepts.
2. Additional equipment. Proper injection and safety devices are essential, and the grower must be in compliance with these legal requirements.

Fertigation is a new agricultural technology that combines irrigation and fertilization. Fertigation technology uses a pressure system (or a height difference in the terrain), to combine soluble solid or liquid fertilizers with irrigation water, according to the soil nutrient content and the nutritional needs of the crop, and through the help of a controllable pipeline system (Fig. 7.3). When the fertilizer has been integrated into the irrigation water, it then evenly, regularly, and quantitatively waters

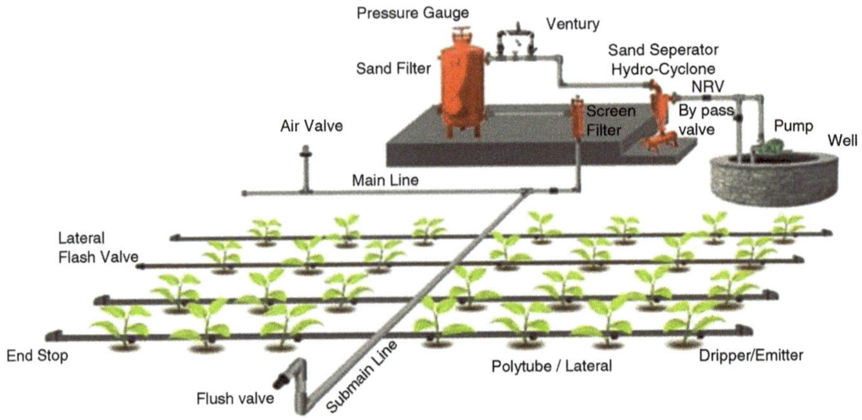

Fig. 7.3 Fertigation system

the growth and developmental root zone of the crops, through the pipeline and the drippers. This allows the root soil to remain loose and to maintain the appropriate water content. It also allows farmers to design the fertigation system according to the characteristics of different vegetables, the soil environment, the current nutritional state of the crops, the water demand of vegetables in different growth periods, and the fertilization cycles, so that the crops receive a proportional amount of water and nutrition regularly and quantitatively.

Advantages of Fertigation

1. Nutrients can be applied any time during the growing season according to the crop's needs. Nutrients can be applied uniformly over the field if the irrigation system distributes water uniformly.
2. Groundwater contamination is less likely with fertigation because less fertilizer is applied at any given time and the application can correspond to the peak crop requirements.

Disadvantages of Fertigation

1. Fertilizer distribution is only as uniform as the irrigation water distribution. Use pressure gauges to ensure that the center pivot is properly pressured.
2. Lower-cost fertilizer materials such as anhydrous ammonia often cannot be used.
3. Fertilizer placement cannot be localized, as in banding.
4. Many fertilizer solutions are corrosive. Chemigation injection pumps and fittings constructed of cast iron, aluminum, stainless, and some forms of plastic are less subject to corrosion and failure.

7.9 Benefits of SSNM

1. *Higher profits:* SSNM can increase and maintain yields by optimizing the balance between supply and demand of nutrients and providing more balanced plant nutrition (Wang et al. 2007). In general, it improves nutrient use efficiency and provides greater returns on investments in fertilizer (Ortiz Monasterio and Raun 2007).
2. *Reduced nitrous oxide emissions:* Agriculture contributes 70–90% of nitrous oxide (N2O) emissions, mostly from N fertilizer. SSNM reduces N2O emissions by reducing total N application and/or timing applications to crop needs, thus avoiding N losses to volatilization, leaching, and runoff.
3. *Improved disease resistance:* The more balanced NPK nutrition that comes with SSNM may lead to improved resistance to plant diseases (Pasuquin et al. 2014).

7.10 Challenges to Adoption of SSNM

1. Technology and knowledge requirements: SSNM requires knowledge of underlying soil properties and the ability to monitor crops' nutrient status and adjust fertilizer inputs accordingly. While the need to conduct on-farm nutrient trials and soil tests has historically been a barrier to implementation of SSNM, the development of decision support systems and farmer-friendly tools and techniques that use proxy information to calculate nutrient requirements makes SSNM more accessible to farmers and farm advisors.
2. Availability of fertilizers: Cost and access to fertilizers – whether synthetic or organic – are not universal. The development of input markets or identification of on-farm nutrient sources may be a necessary precursor to adoption of SSNM, though SSNM can help farmers make best use of limited nutrient resources.
3. Variable economic benefit: For SSNM to increase farmers' profits, SSNM must deliver either (a) savings from reduced fertilizer use without a reduction in yields or (b) yield increases that are valued higher than the costs of acquiring and using SSNM technology. Farmers are more likely to see positive net returns with high-value crops, where yield increases can substantially increase profits, or when fertilizer prices are high.

7.11 Conclusions

The approach of SSNM involves science-based principles for guiding the judicious and efficient application of fertilizers when needed by crops. It recognizes the inherent spatial variability associated with fields during crop production and provides guidelines for optimal use of indigenous nutrients originating from soil, plant residues, manures, and irrigation water. Thus, SSNM considers indigenous nutrient supply of the soil and productivity targets capable of sustaining higher yields on one hand and assured restoration of soil fertility on the other.

References

Buresh, R. J. (2008). Management made easy: A new decision-making tool is helping rice farmers optimize their use of nutrient inputs. *International Rice Research Institute, Rice Today*.

Cassman, K. G., Dobermann, A., Cruz, P. S., Gines, G. C., Samson, M. I., Descalsota, J. P., Alcantara, J. M., Dizon, M. A., & Olk, D. C. (1996). Soil organic matter and the indigenous nitrogen supply of intensive irrigated rice systems in the tropics. *Plant and Soil, 182*(2), 267–278.

Crain, J., Ortiz-Monasterio, I., & Raun, B. (2012). Evaluation of a reduced cost active NDVI sensor for crop nutrient management. *Journal of Sensors, 2012*, 1–10.

Janssen, B. H., Guiking, F. C. T., Van der Eijk, D., Smaling, E. M. A., Wolf, J., & Van Reuler, H. (1990). A system for quantitative evaluation of the fertility of tropical soils (QUEFTS). *Geoderma, 46*(4), 299–318.

Ortiz-Monasterio, J. I., & Raun, W. (2007). Reduced nitrogen and improved farm income for irrigated spring wheat in the Yaqui Valley, Mexico, using sensor based nitrogen management. *The Journal of Agricultural Science, 145*(3), 215.

Pagani, A., Sawyer, J. E., & Mallarino, A., (2013). Site-specific nutrient management: For nutrient management planning to improve crop production, environmental quality, and economic return.

Pasuquin, J. M., Pampolino, M. F., Witt, C., Dobermann, A., Oberthür, T., Fisher, M. J., & Inubushi, K. (2014). Closing yield gaps in maize production in Southeast Asia through site-specific nutrient management. *Field Crops Research, 156*, 219–230.

Portch, S., & Hunter, A. (2002). A systematic approach to soil fertility evaluation and improvement. Special publication, (5), p. 62.

Portch, S., & Stauffer, M. D. (2005). 04 Soil testing: A proven diagnostic tool.

Wang, G., Zhang, Q. C., Witt, C., & Buresh, R. J. (2007). Opportunities for yield increases and environmental benefits through sitespecific nutrient management in rice systems of Zhejiang province, China. *Agricultural Systems, 94*, 801–806. https://doi.org/10.1016/j.agsy.2006.11.006.

Chapter 8
Precision Water Management

Abstract In this chapter differed methods of precision water management in context of agriculture have been described. Agricultural cropping systems depend on the use of water resources for survival, and water needs vary spatially in fields because of spatial soil variability (texture, topography, water holding capacity and infiltration, and drainage rate). Therefore, the need for irrigation may differ between different zones of a particular field. While moving irrigation systems apply water at constant rates, some areas of the field may receive too much water and others not enough. Precision water management (PWM) is related to the judicious use of water so as to achieve sustainable management of water. It refers to the precise application of quality water at the right time, right place, and right stage of crop growth but uniformly across the selected area. Various strategies have been put forward to achieve the same with the help of many technologies and tools. Also, many computer-aided tools have been already developed with the aim to improve water management of irrigation projects. Water is one of the inputs critical to crop growth and yield. Excess and deficit water conditions both can cause stress to the crop. Hence it becomes very necessary to provide with the optimum amount of water to the crop so as to accelerate its yield potential. Quality of the water is another important parameter for the smooth functioning of the crop. Hence, precision water management is defined by both good water quality and adequate amount of water.

Keywords Variable rate irrigation · Soil-landscape water management · Drainage

8.1 Agricultural Water Management

Water is a vital component of agricultural production. It is essential to maximize both yield and quality. Water has to be applied in the right amounts at the right time in order to achieve the right crop result. At the same time, the application of water should avoid waste of a valuable resource and be in sympathy with the environment as a whole. Understanding, measuring, and assessing how water flows around the farm and recognizing how farming practices affect flows will help farmers to manage water efficiently and reduce pollution risks.

L. Ahmad, S. S. Mahdi, *Satellite Farming*,
https://doi.org/10.1007/978-3-030-03448-1_8

The main challenge confronting water management in agriculture is to improve water use efficiency and its sustainability. This can be achieved through the following:

1. An increase in crop water productivity (an increased in marketable crop yield per unit of water transpired) through irrigation
2. A decrease in water losses through soil evaporation that could otherwise be used by plants for their growth
3. An increase in soil water storage within the plant rooting zone through better soil and water management practices at farm and area-wide (catchment) scales

8.2 Precision Water Management

The traditional meaning of precision water management has been that of applying precise amounts of water to crops at precise location and at precise times but uniformly across the field. Precision irrigation is best viewed as a management approach. In most crops growth can proceed and crop yield can be maximized only when the soil moisture potential remains high (and water remains readily available) continuously throughout the growing season.

A review of the rather sparse literature on precision water management brings up a range of definitions, including:

The accurate and precise application of water to meet the specific requirements of individual plants or management units and minimize adverse environmental impact (Raine et al. 2007).

The application of water to a given site in a volume and at a time needed for optimum crop production, profitability or other management objective at the site (Camp et al. 2006).

Applying water in the right place with the right amount (Al-Karadsheh et al. 2002).

Irrigation management (depth, timing) based on crop need to defined sub-areas of a field referred to as management zones (King et al. 2006).

Water is critical to crop productivity since crop yield can be maximized only when the soil moisture potential remains high and crop yields generally increase linearly with water transpired by the crop. Excess water can induce nutrient and aeration stresses and encourage pests that reduce yields and quality. Water management is also critical to water quality, because techniques to optimize water relations for plants can also impact fate and transport of pollutants to surface water. Naturally, the adequacy of water for plant growth is primarily related to the amount, frequency, and distribution of rainfall and soil properties as they affect processes that regulate the hydrological cycle within a watershed. Three approaches of precision water management are:

1. Variable rate irrigation
2. Matching agronomic inputs to water availability defined by soil and/or landscape properties
3. Drainage

Aside from the technology, a major advancement of precision agriculture comes from a change in the way farmers think about their fields. In the past, a field was viewed as a uniform unit: if it's time to water, the entire field gets watered. But the reality is that not every part of the field has the same need for water or any other agricultural inputs. In precision farming terms, these types of applications of agricultural inputs are termed as wastes of resources. Thus, the smartness of precision agriculture tools comes in the form of what is called variable rate technologies that control delivery of water and chemicals according to what that subregion of the field needs by applying the right amount in the right place at the right time. When used to its fullest extent, precision technology reduces a farmer's fertilizer and water consumption – two essential components of farming. Due to the lack of micro-knowledge, farmers have had to resort to blanket applications of these two substances because they lack the necessary technology to pinpoint the smaller areas of real need.

8.3 Variable Rate Irrigation/Precision Irrigation

Hillel (1990) defines a well-managed irrigation system as one that optimizes the spatial and temporal distribution of water so as to promote crop growth and yield as well as to enhance the economic efficiency of crop production (maximum net return). Since the physical circumstances and the socioeconomic conditions for irrigation are site-specific and often season-specific in each case, there can be no single solution to the problem how best to develop and manage the irrigation.

The structure for establishing a strategy for precision irrigation is shown in Fig. 8.1. The first information needed to delineation of in-field spatial variability is originated from the soil-yard maps, which can be found in agricultural planning

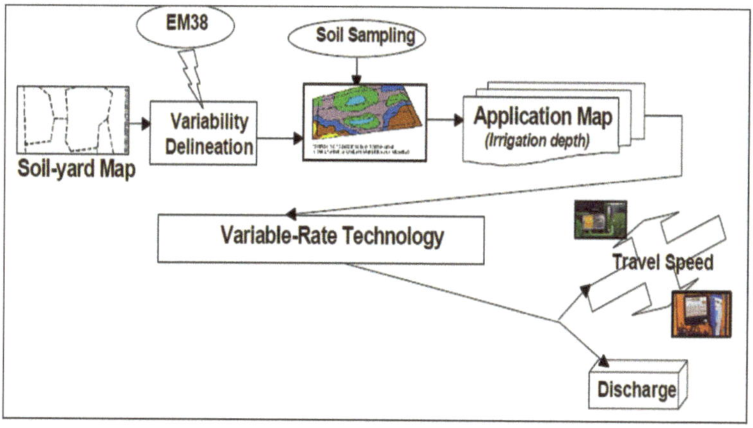

Fig. 8.1 Strategy for precision irrigation

offices. These information are not sufficient to be used for precision irrigation, since these maps provide information about in region (large scale) spatial variability.

The next step is to obtain in-field information (small scale), which comes from taking the advantages of the fast, nondestructive real-time sensors and the concept of surrogated properties, such as electrical conductivity (EC). This must be followed by soil sampling, based on the maps produced from this sensor, and correlate the surrogate property with the property in question (EC vs. AWC). Map for the management zones within the field (application map) for the field activity, here irrigation, showing the different quantities (depths) and their location within the field is established. Then a decision must be taken concerning the technologies that must be integrated with the present field machinery or need to be introduced, here, variable rate technologies. Evaluations for the parameters of this technology, here travel speed and discharge rate, should be done.

The key to a successful precision irrigation system will be defining site-specific goals or objectives, which might include:

- Saving water and reducing costs by applying only the optimum amount of irrigation
- Minimizing adverse environmental impact and better management of the resource base
- Optimizing the economic value of the water applied through irrigation
- Optimizing crop production (yield quantity and/or quality)

8.4 Soil-Landscape Water Management

The potential for precise management of agronomic inputs increases with spatial variability in water availability within the field. Differences in water availability within a field are governed by:

1. The occurrence of dissimilar soil types
2. The presence of soil degradation processes (erosion, compaction, salinization)
3. Variation in landscape

The evidence for spatial variation in water availability is clear. The north-facing slopes had 20% more available water in soils than south-facing slopes throughout the year, whereas soils on east-facing slopes were the driest (Hanna et al. 1982). Crop yields are often highest in the lower slope positions where soil water and nutrient contents are higher. Eroded soils often have lower infiltration rates and lower available water than their non-eroded counterparts. Some portion of landscape variability can be attributed to the variation of soil properties with landscape position, whereas some is attributable to redistribution of water within a landscape due to either runoff or subsurface horizontal flow of water. Compacted soils reduce infiltration or restrict plant roots, thereby limiting water availability to plants. Areas of high salinity are known to reduce yields.

Knowledge of the spatial distribution of water availability can be used as a basis for site-specific input recommendations. There are three approaches for mapping soil water variability

1. County soil surveys
2. Interpolation of a network, usually a grid, of point samples to estimate spatial distribution of soil properties or water content
3. Soil-landscape models to estimate spatial patterns of soil water availability

The presence of small-scale spatial variability in soil physical properties and the high cost of network sampling may limit its use in mapping water availability. Site-specific soil water monitoring is used to some extent as a basis for variable rate irrigation and land in landscape studies. Statistical models of soil-landscape relationships offer opportunities to map spatial patterns of soil properties where relief or some landscape attribute is a primary factor contributing to soil variability. Soil-landscape models are important because terrain modifies the distribution of hydrologic and erosional processes (i.e., soil water content, runoff, and sedimentation), and soil temperature in fields, all important is regulating crop productivity of topography with a regular grid of elevation observations, is referred to as a digital terrain model (DTM) when attributes of a landscape are of interest and a digital elevation model (DEM) when merely relief is represented. A DTM allows the estimation of derivatives of elevation including slope, curvature, aspect, catchment area, and surface drainage proximity variables that correlate to soil and land qualities.

The value of DTM is that it increases the resolution of soil maps for use in site-specific management and in environmental modeling by using terrain attributes to spatially distribute estimated soil attribute data. Therefore, terrain modeling efforts have focused on its application to soil survey to model and depict the spatial variability of soil horizons in reference to the topographic surface and spatial application of simulation models to evaluate current and potential management practices regarding their effects on crop production and the environment in space and time. The extent of use of soil-landscape models is currently limited in-field applications of precision agriculture. However, high-resolution DEMs can easily be created using DGPs and laser-based systems with high vertical accuracies. As elevation maps become available, soil-landscape modeling techniques such as DTM will be increasingly used in precision agriculture.

8.5 Drainage

Poor drainage is often cited by farmers as a source of yield variability within field. Many options for drainage currently exist and can be applied site-specifically. Therefore, there is little need to design site-specific drainage practices. The decision to install drainage is economically, not technically, limited, regardless of scale, the decision to drain hinges on the expectation of returns on investment that exceed costs of installation. Site-specifically, the cost of draining portions of fields or small isolated areas may be higher because the yield depression due to poor drainage can be accurately assessed if sufficient years are included in the calculation. Drainage, therefore, is a site-specific, economic decision based on the conditions at each site and cannot be generalized.

8.6 Benefits of Precision Irrigation

Precision irrigation has the potential to increase both the water use and economic efficiencies. It has been reported that precision irrigation (drip and sprinkler) can improve application efficiency of water up to the tune of 80–90% as against 40–45% in surface irrigation method (Dukes and Scholberg 2004). Results from case studies of variable rate irrigation show water savings in individual years ranging from 0 to 50%. The potential economic benefit of precision irrigation lies in reducing the cost of inputs or increasing yield for the same inputs (Shah and Das 2012).

1. *Water savings*

The primary goal of precision irrigation is to apply an optimum amount of irrigation throughout fields. It is reported by many researchers as the most likely means of achieving significant water savings (Evans and Sadler 2008). The site-specific or variable rate irrigation is considered as a necessary or essential component of precision irrigation. It has been reported that variable rate irrigation could save 10–15% of water used in conventional irrigation practice (Yule et al. 2008). Hedley and Yule (2009) suggested water savings of around 25% are possible through improvements in application efficiency obtained by spatially varied irrigation applications.

2. *Yield and profit*

The experimental studies were carried out by King et al. (2006) for measuring the yield of potatoes under spatially varied irrigation applications. It was reported that yields were better in 2 consecutive years over uniform irrigation management. Booker et al. (2006) analyzed yields and water use efficiency for spatially varied irrigation over 4 years for cotton. They concluded that cotton seems to be unpredictable to manage with spatially varied irrigation. This result is supported by the work of Bronson et al. (2006).

8.7 Components of Precision Irrigation System

(a) *Data acquisition*

A precision irrigation system requires the ability to identify and quantify the variability, i.e., spatial and temporal variability, that exists in soil and crop conditions within a field and between fields. Existing technology is available to measure the various components of the soil–crop–atmosphere continuum many in real time so as to provide precise and/or real-time control of irrigation applications.

(b) *Interpretation*

Data has to be collected, interpreted, and analyzed at an appropriate scale and frequency. The inadequate development of decision support systems has been identified as a major bottle neck for the interpretation of real-time data and adoption of precision agriculture (McBratney et al. 2005).

(c) *Control*

The ability to optimize the inputs and adjust irrigation management at appropriate temporal and spatial scales is an essential component of a precision irrigation system. Applying differential depths of water over a field will be dependent on the irrigation system. Automatic controllers with real-time data should provide the most reliable and accurate means of controlling irrigation applications.

8.8 Conclusions

Precision farming aims to manage production inputs over many small management zones rather than on large zones. Site-specific irrigation management can potentially improve the overall water management in comparison to irrigated areas of hundreds of hectares. A critical element of the irrigation scheduling and management is the accurate estimation of irrigation supplies and its proper allocation for the irrigation offtake structures based on the actual planted areas. The amount of water applied is determined by using a criterion to determine irrigation need and a strategy to prescribe how much water to apply in any situation. The right amount of daily irrigation supply and monitoring at the right time within the discrete irrigation unit is essential to improve the irrigation water management of a scheme.

References

Al-Karadsheh, E., Sourell, H., & Krause, R. (2002, October). Precision Irrigation: New strategy irrigation water management. In *International Conference on Agricultural Research for Development*. Witzenhausen: Deutscher Tropentag.

Booker, J. D., Bordovsky, J., Lascano, R. J., & Segarra, E. (2006, January). Variable rate irrigation on cotton lint yield and fiber quality. In *Beltwide Cotton Conferences* (p. 1768). San Antonio.

Bronson, K. F., Booker, J. D., Bordovsky, J. P., Keeling, J. W., Wheeler, T. A., Boman, R. K., Parajulee, M. N., Segarra, E., & Nichols, R. L. (2006). Site-specific irrigation and nitrogen management for cotton production in the Southern High Plains. *Agronomy Journal, 98*(1), 212–219.

Camp, C. R., Sadler, E. J., & Evans, R. G. (2006). Precision water management: current realities, possibilities, and trends. In *Handbook of precision agriculture* (pp. 153–183). London: Food Products Press.

Dukes, M. D., & Scholberg, J. M. (2004). Automated subsurface drip irrigation based on soil moisture. In *2004 ASAE Annual Meeting* (p. 1). American Society of Agricultural and Biological Engineers.

Evans, R. G., & Sadler, E. J. (2008). Methods and technologies to improve efficiency of water use. *Water Resources Research, 44*(7).

Hanna, A. Y., Harlan, P. W., & Lewis, D. (1982). Soil available water as influenced by landscape position and aspect. *Agronomy Journal, 74*(6), 999–1004.

Hedley, C. B., & Yule, I. J. (2009). Soil water status mapping and two variable-rate irrigation scenarios. *Precision Agriculture, 10*(4), 342–355.

Hillel, D. (1990). Role of irrigation in agricultural systems. *Agronomy*, (30), 5–30.

King, B. A., Stark, J. C., & Wall, R. W. (2006). Comparison of site-specific and conventional uniform irrigation management for potatoes. *Applied Engineering in Agriculture, 22*(5), 677–688.

McBratney, A., Whelan, B., Ancev, T., & Bouma, J. (2005). Future directions of precision agriculture. *Precision Agriculture, 6*(1), 7–23.

Raine, S. R., Meyer, W. S., Rassam, D. W., Hutson, J. L., & Cook, F. J. (2007). Soil–water and solute movement under precision irrigation: knowledge gaps for managing sustainable root zones. *Irrigation Science, 26*(1), 91–100.

Shah, N. G., & Das, I. (2012). Precision irrigation sensor network based irrigation. In *Problems, Perspectives and Challenges of Agricultural Water Management* (pp. 217–232). Rijeka: InTech.

Yule, I. J., Hedley, C. B., & Bradbury, S. (2008, September). Variable-rate irrigation. In 12th Annual Symposium on Precision Agriculture Research & Application in Australasia. Sydney.

Chapter 9
Precision Pest Management

Abstract Pests associated with agricultural production include weeds, insects, and diseases. Most problems faced in agriculture because of pesticide use have to do mainly with groundwater and surface water quality. It is documented that a degree of spatial variation exists in weed distribution. This is mainly because of organic matter, soil texture, landscape position, and the interaction of these factors with crop management crop cultivars, tillage planting density, cultivation, and herbicide application methods. Precision agriculture provides an enabling set of technologies to help reduce potential environmental problems from pest management. The technologies include field maps of weed distribution throughout the year, detection methods for weeds within a field, application methods to apply herbicides only on selected area, and the indications of the effects of weed population on crop yields with data from yield monitors on combines. Each of these technologies allow for the adoption of precision agriculture to weed management. Insects and diseases can be treated similar to weeds using the same principle. The chapter gives details of precision pest and weed management in agriculture. Different principles and practices associated with the precise management of pests and weeds have been discussed.

Keywords Precision pest management · Integrated pest management · Integrated weed management · Weed control

9.1 Integrated Pest Management

The concept of "pest management" was proposed in 1961 (Geier and Clark 1961). Integrated pest management (IPM) is a recently developed technology for pest control that is aimed at achieving desired control while reducing the use of pesticides. Integrated pest management (IPM) can help protect water quality by minimizing the amount of pesticides that producers use and by helping producers to apply pesticides in ways that decrease the risk of chemicals washing off fields into lakes and rivers or leaching into ground water. The high-intensity IPM practice provides an opportunity for the producer to develop multiple management strategies that will

L. Ahmad, S. S. Mahdi, *Satellite Farming*,
https://doi.org/10.1007/978-3-030-03448-1_9

integrate all aspects of pest management within the agriculture production system – this is called integrated pest management, or IPM.

It may also be defined as "A site-specific combination of pest prevention, pest avoidance, pest monitoring, and pest suppression strategies."

9.2 Principles of Integrated Pest Management

IPM methods are performed in three stages: *prevention*, *observation*, and *intervention*. It is an ecological approach with a main goal of significantly reducing or eliminating the use of pesticides while at the same time managing pest populations at an acceptable level.

An IPM system is designed around six basic principles:

1. Acceptable pest levels

The emphasis is on control, not eradication. IPM holds that wiping out an entire pest population is often impossible, and the attempt can be economically expensive and environmentally unsafe. IPM programs first work to establish acceptable pest levels, called action thresholds, and apply controls if these thresholds are crossed. These thresholds are pest and site specific, meaning that it may be acceptable at one site but at another site it may not be acceptable. By allowing a pest population to survive at a reasonable threshold, selection pressure is reduced. This stops the pest gaining resistance to chemicals produced by the plant or applied to the crops. If many of the pests are killed, then any that have resistance to the chemical will form the genetic basis of the future, more resistant, population. By not killing all the pests, there are some unresistant pests left that will dilute any resistant genes that appear.

2. Preventive cultural practices

Selecting varieties best for local growing conditions, and maintaining healthy crops, is the first line of defense, together with plant quarantine and "cultural techniques" such as crop sanitation (e.g., removal of diseased plants to prevent spread of infection).

3. Monitoring

Regular observation is the cornerstone of IPM. Observation is broken into two steps: first, inspection and, second, identification. Visual inspection, insect and spore traps, and other measurement methods and monitoring tools are used to monitor pest levels. Accurate pest identification is critical to a successful IPM program. Record-keeping is essential, as is a thorough knowledge of the behavior and reproductive cycles of target pests. Since insects are cold-blooded, their physical development is dependent on the temperature of their environment. Many insects have had their development cycles modeled in terms of degree days. Monitor the degree days of an environment to determine when is the optimal time for a specific insect's outbreak.

4. Mechanical controls

They include simple hand-picking, erecting insect barriers, using traps, vacuuming, and tillage to disrupt breeding.

5. Biological controls

Natural biological processes and materials can provide control, with minimal environmental impact, and often at low cost. The main focus here is on promoting beneficial insects that eat target pests. Biological insecticides, derived from naturally occurring microorganisms (e.g., *Bt*, entomopathogenic fungi, and entomopathogenic nematodes), also fit in this category.

6. Responsible pesticide use

Synthetic pesticides are generally only used as required and often only at specific times in a pest's life cycle. Many of the newer pesticide groups are derived from plants or naturally occurring substances (e.g., nicotine, pyrethrum, and insect juvenile hormone analogues), but the toxophore or active component may be altered to provide increased biological activity or stability.

9.3 Steps in Integrated Pest Management

The IPM philosophy of pest management involves three fundamental steps:

1. Use cultural methods, biological controls, and other alternatives to conventional chemical pesticides when practical.
2. Use field scouting, pest forecasting, and economic thresholds to ensure that pesticides are only used against real and not perceived pest problems.
3. Match pesticides with field site features so that the risk of contaminating water is minimized. Substitute lower-risk pesticides when feasible, and alternate the use of pesticides from different chemical classes.

1. Nonchemical alternatives

Overreliance on any single pest control measure can have undesirable effects. Pesticides can contribute to pest outbreaks by eliminating natural enemies and allowing pests to rebound without checks. Cultural methods are those good farming practices that make the environment less suitable for pest colonization and survival. Biological control involves using predatory, parasitic, and disease-causing organisms for insect pest control as well as using competitive or antagonistic organisms for weed suppression. It also includes conservation of naturally occurring beneficial insects.

2. Scouting

Crop scouting is a crucial component in any IPM program used to identify the pest (Fig. 9.1). The effectiveness of both proactive and reactive pest management

Fig. 9.1 Crop scouting

measures depend on correct identification. The scouting must be done by a person trained in IPM. Proper monitoring (scouting) can determine pest population levels and locations within the field. Pests should be suppressed at a point when their levels reach a particular point known as the economic threshold. It is the point where the cost to control the pest equals the crop damage caused by the pest. Controlling a pest prior to this level is therefore not cost-effective.

Crop scouts move methodically through fields looking for pests, measuring plant and pest populations, and collecting information that growers can then use to make important pest control decisions. This information can help growers know not only if pest control measures are needed but where they are needed.

3. Pest forecasting

Pest forecasting uses information or data to predict pest problems early. In precision farming systems, insect-pest forecasting provides opportunity to the farmers for preparedness and takes timely action to apply bioagents and pesticides, which ultimately cut down the cost of production and serve as an important tool in agriculture. Pest forecasting models are essential tools and prerequisite of IP1 as the insect population levels are deciding factors in pest management. In India, one of the earliest works on pest forecasting was started by Dr. S. Pradhan when he gave the concept of biotic theory of locust population explosion. According to him, temperature was one of the most important factors, which determines the duration of various stages of insect life cycle and the number of generations. Remote sensing has also been used for forecasting of locusts (*S. gregaria*) in India. The locusts oviposit in the sandy deserts where moisture is present in the soil. The satellite detection of moist soil sand as breeding site and the amount of green vegetation as food in the desert give the possible amount of locust population. The migration of locusts is forecasted using the wind direction (Mahal et al. 2012).

In addition, models have been developed, like the degree-day approach, which can help determine when scouting should begin, or when pesticide application will have the maximum control.

4. Record-keeping

Records are an important tool to track pest populations over time and can document reduction in pesticide use. An IPM plan must be developed and be provided to NRCS. Nonchemical pest management practices must be recorded. Documentation shall include target pest, method or technique used, date, and/or crop stage when used. All pesticide use must be recorded. Documentation shall include product name or active ingredient, application location (field identification), target pest, application rate, application timing, and extent of application (entire field vs. spot treatment, for example).

9.4 Integrated Weed Management

Integrated weed management (IWM) means integrating multiple methods to manage weeds, using the combination of practices that is most effective for solving the specific weed issue at hand. These weed management techniques form a "toolbox" in which each "tool" can be integrated into a weed management plan catered to the particular farm and problem. The toolbox includes chemical (herbicide), mechanical, cultural, biological practices, and prevention of weed introduction and spread.

IWM tactics span a wide range of types and complexity. Not all IWM tactics are very complex. Some examples include equipment cleaning, timely scouting, altering herbicide tank mixes, rotating herbicides, cover cropping, changing tillage practices, and hand-pulling weeds.

Integrated weed management is not an alternative to herbicides in conventional crops. For many decades, herbicides have been the primary means of weed management in conventional crops due to their simplicity, effectiveness, and affordability. IWM is about using all options available to best solve the problem – in many cases in conventional crops, herbicides are part of this solution.

9.5 Weed Control and Weed Management

Weed control and weed management are not synonymous. Weed control refers to actions used to achieve the immediate elimination of an existing weed population, usually through the use of herbicides and tillage. Weed management goes beyond control of existing weed problems and focuses on reducing weed invasion and emergence, preventing weed reproduction, and minimizing weed competition with crops. Thus, while weed control reacts to problems after they occur, weed management emphasizes understanding the causes of weed problems with the goal of preventing weeds from becoming problematic in the first place.

9.6 Tactics of Integrated Weed Management

Integrated weed management (IWM) combines the use of biological, cultural, mechanical, and chemical practices (Fig. 9.2) to manage weeds, so that reliance on any one weed management technique is reduced. The main goals of an IWM program are to:

1. Use preventive tools to maintain weed density at a level that does not harm the crop
2. Prevent shifts toward more difficult to control weeds
3. Develop agricultural systems that maintain or improve crop productivity, farm revenues, and environmental quality

Thus, designing a successful IWM program requires understanding the different biological and ecological factors that influence the short-, mid-, and long-term dynamics of weeds in agricultural settings.

1. Prevention

Monitor inputs to the farm to avoid bringing in things that may be contaminated with weed seeds. To do this, learn how key weeds are spread and whether those weeds are located in areas that the farm is transporting supplies from. Equipment,

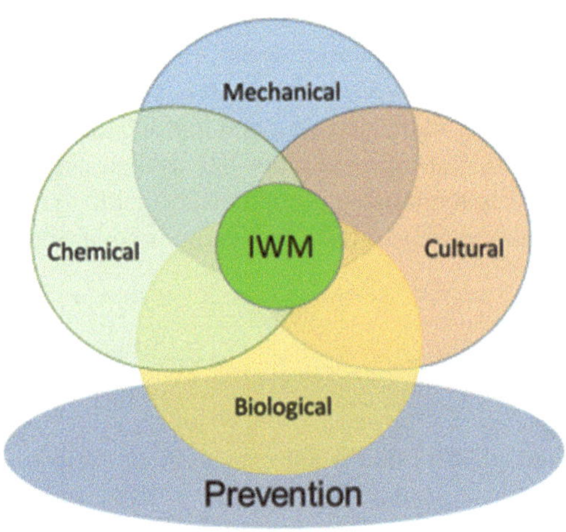

Fig. 9.2 Five types of management tactics used in integrated weed management

manure, feed, and crop seed are primary spreaders of weeds. Spread via wind and wildlife is more common for some weed species than others.

2. Chemical

Herbicides are a key part of IWM in conventional and some organic systems. In conventional crops, using multiple effective herbicide modes of action (MOA) is essential for effective control of resistant weeds. This involves combining multiple MOA in tank mixes and varying MOA between applications and seasons. For MOA with high occurrences of resistant weeds, avoid repeat use in consecutive seasons.

3. Cultural

Cultural tactics are crop management decisions that help the crop be more competitive against weeds and help optimize the effectiveness of herbicide applications. Common examples include timely scouting, row spacing, crop rotation, crop variety selection, timing of planting, and cover cropping. Information about using these tactics for weed management is found throughout this site.

4. Mechanical

Common mechanical tools to disrupt weed growth and survival include cultivation, tillage, burning, and hand-weeding. Mechanical IWM tools also include emerging technologies like harvest-time seed destructors, cover crop rollers, and robotic weeders. Mechanical tools should be integrated when appropriate as part of a larger IWM program. Many of these mechanical techniques are available to no-till growers.

5. Biological

A less common IWM strategy is the use of living organisms, including livestock, insects, nematodes, fungi, and bacteria, to target weeds. Many biological agents target specific weed species, while livestock are relatively more generalist in weed consumption and may avoid eating certain weeds.

9.7 Advantages of IPM/IWM

Besides lowering the impact of chemical substances on the biota in the ecosystem, there are several other benefits of IPM/IWM, such as:

1. Slower development of resistance to pesticides/herbicides

Pests/weeds can develop a resistance to pesticides/herbicides over time. When the applications of the chemicals are used repeatedly, the pests/weeds can develop a resistance to the pesticides/herbicides via natural selection, where the pests/herbs that survive the application of the chemicals will pass on their genes to their offspring (Bethke 2010). This leads to the creation of "superpests" (Garthwaite 2012). IPM reduces the risk of this occurring as the methods adopted by IPM/IWM are natural.

2. Maintaining a balanced ecosystem

The use of pesticides may eradicate the pest/weed population. However, there is a risk that nontarget organisms/plants are also affected, which can result in species loss. IPM/IWM can eradicate pests/weeds while maintaining the balance of the ecosystem.

3. Better cost vs. value margin

The reduced usage of pesticides/herbicides is more cost-effective in the long term, as IPM/IWM controls pests/weeds when there are surges, as opposed to the regularly timed application of pesticides/herbicides.

9.8 Disadvantages of IPM/IWM

Disadvantages of IPM/IWM include:

1. More involvement in the technicalities of the method

Individual farmers and all those involved in IPM/IWM have to be educated about their options in the various methods available, which often take time.

2. Time and energy-consuming

Application of IPM/IWM takes time and has to be closely monitored, as the practice of IPM/IWM has many different methods integrated in order to provide the most effective pest control methods. Different pests/weeds have different control methods, and it is necessary to monitor which methods are the best for specific pests/weeds.

However, the disadvantages are easily offset with the establishment of organizations that actually provide training and education to IPM/IWM practitioners.

9.9 Conclusions

Sustainable development is rooted in the concept of ecosystem integrity and permeates all facets of human endeavor, whether economic, social, or cultural. The scope of concerns and practices of IPM correspond closely to that of sustainable development. Each emphasizes preservation of processes associated with natural ecosystems, long-term well-being of humans as members of communities, economic viability, and deployment of exogenous resources only after careful consideration. For agriculture, concepts underlying IPM can be equated with concepts underlying sustainable agriculture. For both, one can expect concepts to evolve further over time.

References

Bethke, J. A. (2010). *UC IPM pest management guidelines: Floriculture and ornamental nurseries*. UC ANR Publication, 3392.

Garthwaite, J. (2012). Superweeds, superpests: The legacy of pesticides in energy, the environment and the bottom line. *The New York Times*.

Geier, P. W., & Clark, L. R. (1961). *An ecological approach to pest control*. In Proceedings of the eighth technical meeting. International Union of Conservation of Nature and Natural Resources, Warsaw, 1960, pp. 10–18.

Mahal, M. S., Singh, B., Sarao, P. S., & Singh, S. (2012). Role of insect-pest forecasting in integrated pest management. In *Theory and practice of integrated pest management*, p. 248.

Chapter 10
Recent Advances in Precision Agriculture

Abstract The rapid technological developments during the last few years have introduced radical changes in the working environment in the agricultural sector. Agriculture has entered a new era in which the key to success is access to timely information and elaborated decision-making. The up-to-date and skilled farm manager has to choose between various production options utilizing the latest advancements in research and technology. Decision-making is an important aspect in farm management. The basis for enhanced decision-making is availability of timely, high-quality data. Technologies now available allow the concept of precision agriculture to be realized in a practical production setting. Some of the new concepts in the technologies have been discussed in this chapter.

Keywords Internet of Things · Cloud computing · Drones · Unmanned aerial vehicles · Cloud agro system · Farm management system

10.1 Internet of Things in Precision Agriculture

The concept of the Internet of Things (IoT) first became popular in 1999, through the Auto-ID Center at MIT and related market analysis publications. IoT is a broad term that describes the interconnection of different daily life objects through the internet. In the concept of IoT every object is connected with each other through a unique identifier so that it can transfer data over the network without a human to the human interaction (Morais et al. 2005). IoT has referred as a network of everyday objects having ubiquitous computing. The ubiquity of the objects has increased by integrating every object with embedded system for interaction (Coetzee and Eksteen 2011). The aim of IoT is to connect every person and every object through the internet (Shahzadi et al. 2016). The system architecture for IoT in precision agriculture is shown in Fig. 10.1.

The system has useful applications in farm as well as greenhouse. The central system is the heart of this architecture, as it does half of the work of system. Central system is responsible for communications between nodes and central server and database management as well as communication with the outer world. Central

L. Ahmad, S. S. Mahdi, *Satellite Farming*,
https://doi.org/10.1007/978-3-030-03448-1_10

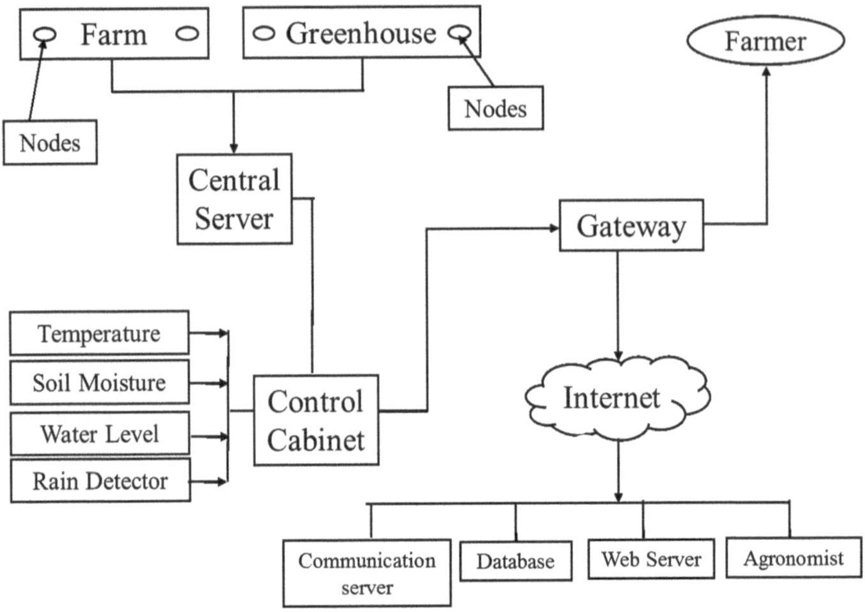

Fig. 10.1 System architecture for IoT

system consists of three main entities which are communication server, database, and Web server. Control cabinet provides the easy access to the nodes, which consists of temperature sensor, soil moisture sensor, water level sensor, and rain detector sensor. The data sensed by the nodes is transferred to the central system by central server. Any warnings or notifications are communicated with farmer via gateway provided. Agronomist is contacted by central system in case of unusual activities (Savale et al. 2015).

10.2 Prerequisites of IoT Applications in Agriculture

The accessible, affordable, and interactive crowdsourcing platform for sustainable agriculture would provide a means for sharing information regarding traditional sustainable agricultural methods, techniques, tools, tips, etc. and allow interactivity and offline data entry for consolidated information upload. Addressing the food security/water security with sustainable agriculture, the solution must provide supplementary information/services such as third-party agricultural, microfinance services, etc. for farmers. It must also provide a centralized repository for a variety of information such as traditional sustainable farming techniques, crop diseases, etc. coming through various sources, allowing interactive farming and easy access to users over various devices such as mobile phones, IVR, computers, and kiosks in addition to providing multilingual support of traditional practices with modern value. The system should meet the following requirements (Patil et al. 2012):

1. *Robust models*: The characteristic features of agriculture sector such as diversity, complexity spatiotemporal variability, and uncertainties have to be considered in developing the right kinds of products and services.
2. *Scalability:* The size of farms varies from small to large, and hence the solutions should be scalable. The testing and deployment happen in stages, and therefore the architecture should be able to scale up incrementally with less overheads.
3. *Affordability:* Affordability is the key to success. The cost has to be appropriate with substantial benefits. Standardized platforms, tools, products, and services can bring the cost down with increased volumes.
4. *Sustainability:* The issue of sustainability is vital because of intense economic pressure and fierce global competition.

10.3 Structure of IOT for Agriculture

The system has three layers, namely, sensor layer, transport layer, and application layer. Their functions are as follows:

1. *Sensor/information collection layer*: The main task of this layer is to achieve automatic and real-time transformation of the physical figures of real-world agricultural production into digital information or data that can be processed in virtual world through various means. The information categorize that Internet of Things for agriculture collects are:
   - Agricultural sensor information: temperature, humidity, pressure, gas concentrations and vital signs, etc.
   - Agricultural products attribute information: name, model, feature and price, etc.
   - Agricultural working status information: operating parameters of apparatus, equipment, etc.
   - Agricultural location information: location of products, etc.

   The main task of information collection layer is to mark the various kinds of information and collect the marked information and the physical information in the real world by sensing techniques and then transforming them to digital information for processing. Information collection layer involves these techniques: two-dimensional code labels and readers, RFID tags and readers, cameras, GPS sensors, terminals, cable networks, sensor networks, and wireless networks.

2. *Transport/network layer*: The main task of this layer is to collect and summarize the agricultural information acquired through sensor layer for processing. Transport layer is the nerve center and cerebra of Internet of Things for agricul-

ture, transmitting and processing data. The network layer includes the integration of the Internet network and telecommunication, network management center, information center, and intelligent processing centers.

3. *Application layer*: The main task of this layer is to analyze and process the information collected so as to cultivate digital awareness of the real word. It is a combination of IOT and Agricultural Market intelligence.

10.4 Cloud Computing

Cloud computing is a tool to make IT related services available in a simplified manner hiding the complexities of those services, without really knowing and getting involved in the technicalities of how and what to do in providing the needed services. The term *cloud computing* is given to this approach because the users do not really need to know who is providing those services and users consider that the services are rendered by the cloud – an unknown to them. The charm of cloud computing is that the services may be availed whenever and wherever needed. It also reduces the cost of availing those services drastically. At the same time, it offers involvement of very less manpower and maintenance of those services. It also makes users free from certain concerns such as buying software, maintaining them up-to-date, maintenance of data, etc. All these issues would be taken care of by Cloud providers. Cloud computing offers various models based on user requirement. The applications of cloud computing in agriculture (Fig. 10.2) can be summarized as:

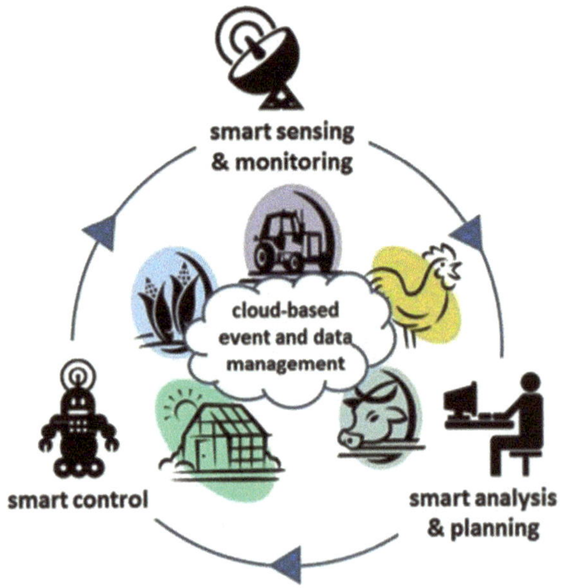

Fig. 10.2 Cloud computing in agriculture

1. Large-scale information storage

Cloud computing provides high-capacity data stores to store the large-scale data and information. The databases pertaining to the farming community such as crop information, weather information, market information, farmers' experiences of the agricultural processes, information about the pesticides and prescriptions, etc. can be easily stored on cloud. The individual farmers can store the information and retrieve the information from cloud easily. Different researchers from agro-related institutions can present the information about the newly developed agricultural techniques and tools to the farming community through cloud. Otherwise cloud is a suitable infrastructure to share the mutual information and experiences among the farmers across the globe. The market condition of different crops-related data is helpful to taking decision in choosing the crops (Sahoo et al. 2013).

2. Low-cost access to IT resources

Cloud computing provides low-cost access to vast IT resources. It implements the pay on usage basis model. The farming community needs not to invest in owning the IT resources, but instead they can access the required resources that can be rented from the cloud. It is the cheaper and reliable method to access the resources. Information communication technology is the modern cloud infrastructure to allocate the resources and services on demand means whenever and whatever a farmer required (Sahoo et al. 2013).

3. Cloud agro system

Cloud agro system is the cloud-based computerized system used to monitor the overall information related to agriculture. The farming community is scattered worldwide where the language and tradition (the ways of carrying out the agro activities) barriers exist. The modern IT tools provides the online language translation mechanisms. Thus the information stored on cloud may be presented to the farmer's native language in the cloud agro system. It helps the farmers to taking decision related to crops production according to the demand and supply of the crops in the market. The researchers can collect the experiences of wide range of farmers using sophisticated IT tools such as online from collection, etc. and experiment the new tools and techniques in the agriculture based on the collected information (Patel and Patel 2013).

4. Easy solution to farming queries

Every profession needs time to time expert opinions. Cloud computing has provided the ways for easy solutions to the problems the farmers may face at the different stages of their farming processes starting from tilling to marketing and selling of their products. In the situations when farmers do not have the answers to the queries at their own cloud computing is an easy alternative. They can get online expert advice from the stored databases on cloud. Thus, cloud computing is proved to be channel to solve the farmers difficulties immediately because it respond to the farmers very faster and accurately. Farmers can post regarding any disease and pesticide and may decide according to the expert replies (Patel and Patel 2013).

5. Land record automation

With the availability of large-scale storage infrastructure, the land records are being computerized in the entire word. Cloud computing storage facility store the record of land with the description related to that particular land like soil analysis result, production history, etc. Different corporations store the accurate data pertaining to land records after the proper verifications of facts and figures (Hori et al. 2010).

6. Farm management system (FMS)

Farm management system provides the disease alert system and reaction of the disease on the crops. It also gives the knowledge about the new software-related machinery and techniques applied in agriculture. Therefore, FMS organize online training program for the farmers to aware them about the diseases and pesticides and also tells about how to keep the maintenance of a farm. FMS has three components: knowledge-based preparation, training of the classifier, and classifier update (Kaloxylos et al. 2014).

7. Data collection tools

In the modern time, many effective and reliable data collection tools are available which can be easily integrated with cloud computing applications, for example, wireless network, radio frequency identification sensors, WiMax, internet, etc. Sensors have the applications in measuring quality of water and soil, forestry, predicting changing environmental conditions, etc. The sensors can be soil gravity water detection sensor, a humidity sensor, a chemical sensor for measuring of the pH of liquids, and a bending sensor for gauging pressure (Hori et al. 2010).

8. Weather forecasting

Cloud computing provides the weather forecast for specific duration to the farmers so that they can take decision related to selection of the crops. Sometimes climate is not suitable for seasonal crop in that condition farmer can choose an alternative seasonal crop (Patel and Patel 2013).

10.5 Drones or Unmanned Aerial Vehicles (UAVs)

Drones were first introduced during WWI for enemy surveillance. They are small, radio-controlled airplanes that can be used to take pictures. Nowadays, their use has expanded to small-time hobbyists and is quickly expanding to use in agriculture. Crop scouting drones are small enough to be brought to the field in the back of a pickup. A lot of scouting these days is done by interns on foot.

At ground level it is hard to cover the entire field and get a good look at the bigger picture. Especially late season, when most crops are over a person's head, it is nearly impossible to see the whole field on foot. Also, farms continue to get larger and more acres have to be watched. Once an entire field is covered by a drone, trouble spots can be identified and targeted for scouting on foot.

10.6 Data Collected

The simplest drone systems come with a basic digital camera that only takes still images. A simple camera from can be mounted on a drone and equipped with different filters. Other types of cameras or sensors can be attached to suit many different kinds of uses. Orchard growers look more at the status of the crop instead of focusing on weed pressure like a cash crop farmer. Researchers will use more technical sensors that reach beyond the normal range of simple scouting.

A more advanced camera filter for crop scouting is one that takes near-infrared images. Healthy plants reflect both green and infrared wavelengths of light. When they are stressed from pest, nutrient, or water pressures, the types of light reflected changes and can be picked up by the cameras. On the pictures, healthy plants will appear bright red, while stressed plants or weeds will look darker.

Following are six ways aerial and ground-based drones can be used throughout the crop cycle:

1. *Soil and field analysis*: Drones (Fig. 10.3) can be instrumental at the start of the crop cycle. They produce precise 3-D maps for early soil analysis, useful in planning seed planting patterns. After planting, drone-driven soil analysis provides data for irrigation and nitrogen-level management.
2. *Planting*: Startups have created drone-planting systems that achieve an uptake rate of 75% and decrease planting costs by 85 percent. These systems shoot pods with seeds and plant nutrients into the soil, providing the plant all the nutrients necessary to sustain life.
3. *Crop spraying*: Distance-measuring equipment – ultrasonic echoing and lasers such as those used in the light detection and ranging, or LIDAR, method – enables a drone to adjust altitude as the topography and geography vary and thus avoid collisions. Consequently, drones can scan the ground and spray the correct

Fig. 10.3 A drone equipped with sprayers for spraying chemicals on crops

amount of liquid, modulating distance from the ground and spraying in real time for even coverage (Fig. 10.3). The result is increased efficiency with a reduction of the amount of chemicals penetrating into groundwater. In fact, experts estimate that aerial spraying can be completed up to five times faster with drones than with traditional machinery.

4. *Crop monitoring*: Vast fields and low efficiency in crop monitoring together create farming's largest obstacle. Monitoring challenges are exacerbated by increasingly unpredictable weather conditions, which drive risk and field maintenance costs. Previously, satellite imagery offered the most advanced form of monitoring. But there were drawbacks. Images had to be ordered in advance, could be taken only once a day, and were imprecise. Further, services were extremely costly and the images' quality typically suffered on certain days. Today, time-series animations can show the precise development of a crop and reveal production inefficiencies, enabling better crop management.
5. *Irrigation*: Drones with hyper-spectral, multispectral, or thermal sensors can identify which parts of a field are dry or need improvements. Additionally, once the crop is growing, drones allow the calculation of the vegetation index, which describes the relative density and health of the crop, and show the heat signature and the amount of energy or heat the crop emits.
6. *Health assessment*: It's essential to assess crop health and spot bacterial or fungal infections on trees. By scanning a crop using both visible and near-infrared light, drone-carried devices can identify which plants reflect different amounts of green light and NIR light (Fig. 10.4). This information can produce multispectral images that track changes in plants and indicate their health (Fig. 10.5). A speedy response can save an entire orchard. In addition, as soon as a sickness is discov-

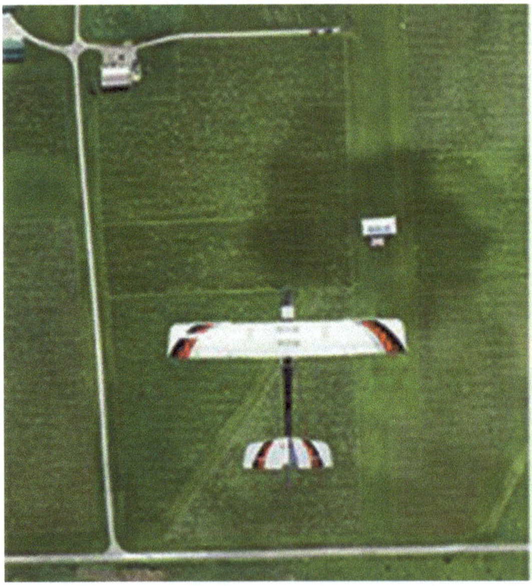

Fig. 10.4 A drone equipped with multiple sensors to image fields

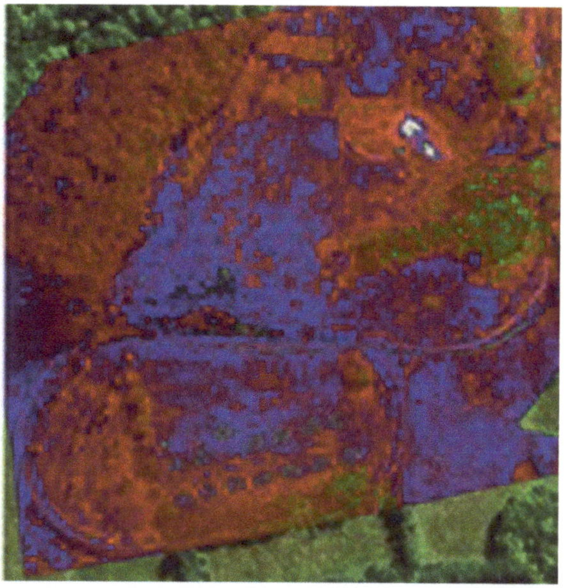

Fig. 10.5 This image depicts vegetation in near-infrared light to show chlorophyll levels

ered, farmers can apply and monitor remedies more precisely. These two possibilities increase a plant's ability to overcome disease. And in the case of crop failure, the farmer will be able to document losses more efficiently for insurance claims.

Drones have other uses than just crop scouting. Livestock ranchers can use them to check up on cattle in a far pasture. Farmers using irrigation equipment can look out for problems without having to walk out on foot. On the business side, real estate agents can use drones to get pictures of property for sale. The cameras can also be used to determine soil types, organic matter levels, and soil moisture levels. The soil maps can be combined with variable rate equipment to adjust the plant population or fertilizer levels based on what the soil can handle. To do this with traditional soil samples would take a large number of samples to get the same resolution across the field, costing time and money. Soil moisture levels are used to adjust irrigation for field conditions depending on weather and wet spots.

10.7 Conclusions

Recent years have seen dramatic changes in agricultural practices, driven largely by efficiency, economic, and environmental considerations. These practices, mainly forms of precision agriculture, have allowed growers to gain a hitherto unprecedented understanding of a crop's condition and requirements and its growing environment, allowing informed decision to be made regarding key functions such as irrigation and the application of fertilizers, herbicides and pesticides. Yields are increased, whereas the costs and environmental impact are reduced.

References

Coetzee, L., & Eksteen, J. (2011, May). *The internet of things-promise for the future? An introduction*. In IST-Africa conference proceedings, 2011 (pp. 1–9). IEEE.

Hori, M., Kawashima, E., & Yamazaki, T. (2010). Application of cloud computing to agriculture and prospects in other fields. *Fujitsu Scientific & Technical Journal, 46*(4), 446–454.

Kaloxylos, A., Groumas, A., Sarris, V., Katsikas, L., Magdalinos, P., Antoniou, E., Politopoulou, Z., Wolfert, S., Brewster, C., Eigenmann, R., & Terol, C. M. (2014). A cloud-based farm management system: Architecture and implementation. *Computers and Electronics in Agriculture, 100*, 168–179.

Morais, R., Valente, A., & Serôdio, C. (2005, July). *A wireless sensor network for smart irrigation and environmental monitoring: a position article*. In 5th European federation for information technology in agriculture, food and environment and 3rd world congress on computers in agriculture and natural resources (EFITA/WCCA) (pp. 845–850).

Patel, R., & Patel, M. (2013). Application of cloud computing in agricultural development of rural India. *International Journal of Computer Science and Information Technologies, 4*(6), 922–926.

Patil, V. C., Al-Gaadi, K. A., Biradar, D. P., & Rangaswamy, M. (2012). Internet of things (Iot) and cloud computing for agriculture: An overview. *Agro Informatics and Precision Agriculture (AIPA 2012)*.

Sahoo, D. G., Mehfuz, D. S., & Rai, R. (2013). Applications of cloud computing for agriculture sector. *CSI Communications*, November, 10.

Savale, O., Managave, A., Ambekar, D., & Sathe, S. (2015, December). Internet of things in precision agriculture using wireless sensor networks. *International Journal of Advanced Engineering & Innovative Technology, 2*(3), 14–17.

Shahzadi, R., Ferzund, J., Tausif, M., & Suryani, M. A. (2016). Internet of things based expert system for smart agriculture. *International Journal of Advanced Computer Science and Applications, 7*(9), 341–350.

Chapter 11
Yield Monitoring and Mapping

Abstract The chapter deals with techniques used for yield monitoring and mapping in precision agriculture. Yield maps are one of the most valuable sources of spatial data for precision agriculture. In developing these maps, it is essential to remove the data points that do not accurately represent the yield at a corresponding location. Map averaging or smoothing is usually done to aid data interpretation. A long yield history is essential to avoid drawing conclusions that are affected by the weather or other unpredictable factors during a particular year. Typically, at least 5 years of yield maps are desired. Processed yield maps can be used to investigate factors affecting the yield or to prescribe variable rate applications of agricultural inputs according to spatially variable yield goals (yield potential). Producers interested in precision farming should, however, always evaluate different management approaches to identify those that provide the greatest benefit at a particular site.

Keywords Yield monitoring · Yield mapping · Yield monitor · Data collection and storage · Calibration · Yield history evaluation

11.1 Yield Monitoring in Precision Agriculture

Yield monitoring is an aspect of precision agriculture that helps to provide farmers with adequate information to make educated decisions about their fields (Wang 1999). Yield monitors are a rather recent development and allow farm equipment such as combine harvesters or tractors to gather a huge amount of information, including grain yield, moisture levels, soil properties, and much more. Due to the fact that yield monitors provide farmers with so much information, they are much more able to assess things such as when to harvest, fertilize, or seed, the effects of weather, and much more.

Yield monitors work in three very simple steps:

1. The grain is harvested and fed into the grain elevator which has sensors that read moisture content of the grain.
2. After that process as the grain is being delivered to the holding tank, more sensors monitor the grain yield.

L. Ahmad, S. S. Mahdi, *Satellite Farming*,
https://doi.org/10.1007/978-3-030-03448-1_11

3. As both of these sensors work, the information is sent to the driver cab and is displayed on a screen as well, and the information is georeferenced so it can be mapped as well as closely investigated on a later time or date.

There are many benefits for farmers to use yield monitoring technology, although one main benefit is that it helps to give the farmer accurate and often georeferenced data about their field. A farmer can better understand crop yield and crop-related information to mitigate potential threats or enhance possible opportunities. Other benefits with a yield monitoring system include the ability for a farmer to export the information onto a personal computer, allowing this information to be available in a variety of different formats, including in equipment displays, at home, or printed. Also, in the home or office, a farmer can use specialized computer software to assess and better understand the recorded information. Yield monitors can monitor grain by a field-by-field or load-by-load basis; this gives the farmer a huge amount of flexibility as well as provides him or her with instant information about the load they have gathered (Thylen and Murphy 1996).

11.2 Components of a Yield Monitor

Yield monitors are a combination of several components (Fig. 11.1). They typically include a data storage device, user interface (display and key pad), and a console located in the combine cab, which controls the integration and interaction of these components. The sensors measure the mass or the volume of grain flow (grain flow

Fig. 11.1 The basic components of a grain yield mapping system

sensors), separator speed, ground speed, grain moisture, and header height. Yield is determined as a product of the various parameters being sensed.

11.2.1 Mass Flow Sensor

A mass flow sensor (Fig. 11.2) is a sensor that helps to provide the yield monitor with enough information to establish a grain yield measurement. The mass flow sensor works by using a load cell which is fixed to the top of a clean grain elevator. When the harvested grain is fed through the combine, it eventually will hit up against this load cell; this is then transformed into an electrical signal and relayed to the yield monitor. The yield monitor uses this reading to determine how much grain is being taken into the combine at any point in time. This sensing method is very common, although there are different methods used, as well as variations of the same method.

11.2.2 Moisture Sensor

Being aware of the moisture content in grain that has been harvested can be extremely valuable information for a farmer to know, especially for aspects including harvesting, storing, and drying crops. When farmers take these moisture readings, they are better able to obtain an accurate market value for their crop. The moisture sensor works when the grain moves in between two conductive surfaces, which measure how much electric charge the grain can store – this is known as capacitance (Fig. 11.3). There are various places to mount the moisture sensor, and it is a vital step of the yield monitoring process.

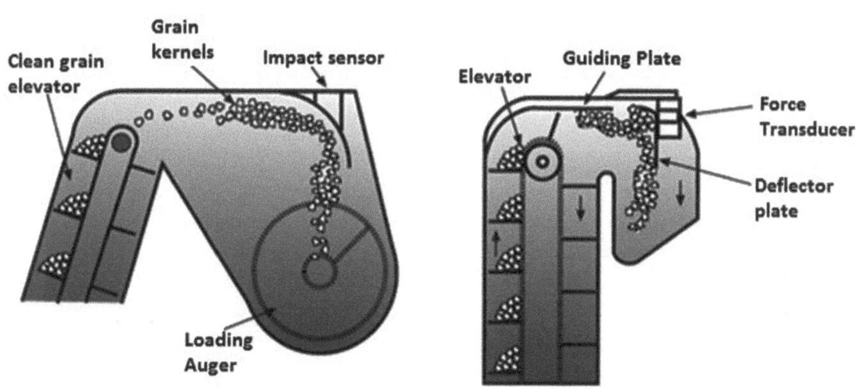

Fig. 11.2 Mass flow sensor on yield monitor

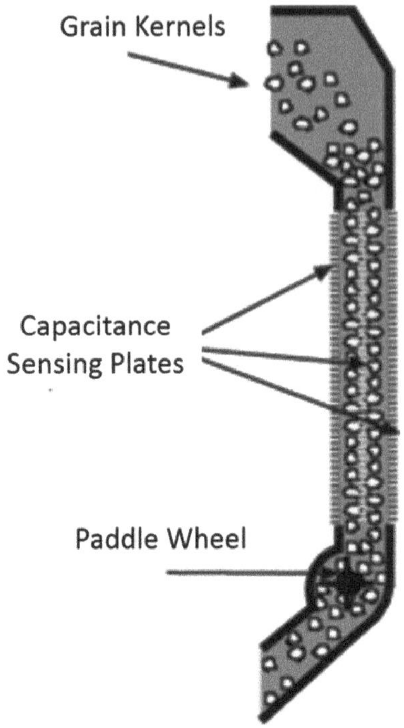

Fig. 11.3 Moisture sensor installation on the clean grain elevator for yield monitoring

11.2.3 GPS Receiver

A GPS receiver, or global positioning system receiver, is a remote sensor that measures a variety of different pieces of data, including where the equipment is located, speed, altitude, and much more. A GPS receiver is a primary component when georeferencing, as the GPS receiver helps to record the position of the equipment in use and then sends that data to an onboard computer which then connects it with all other information that the computer has collected at that particular location. The GPS receiver is one of the key components that can help to transform yield monitoring data from graphs and charts, into tangible maps that the farmer can use.

11.2.4 Yield Monitor

Sometimes referred to as a task computer, receiver, or yield monitor, this piece of technology is the monitor that is located in the cab of the combine or tractor. This piece of equipment serves many different purposes, although its main function is to display the information gathered by the different onboard sensors to allow the operator to know in a real-time manner different moisture levels, crop yield, and more (Fig. 11.4). These monitors also have the capability to store memory as well as

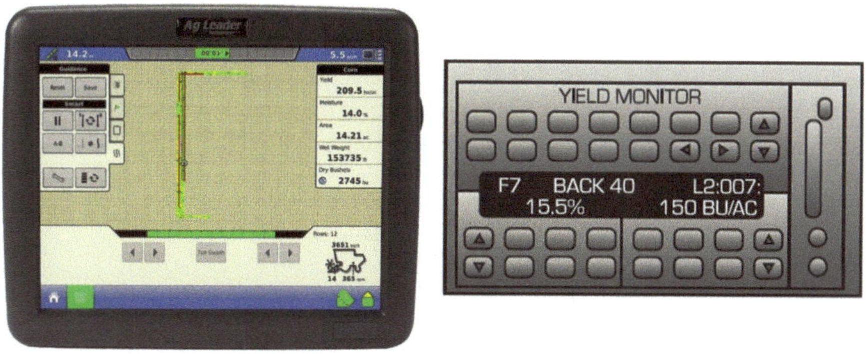

Fig. 11.4 Yield monitor

transfer memory to a laptop or home computer. This transfer of memory makes analyzing data a much more comfortable experience, and it also provides the capability of using more software to interpret and render the data collected by the yield monitoring system.

11.3 Data Collection and Storage

Data are often recorded on removable memory cartridges, such as a Personal Computer Memory Card Industry Association (PCMCIA) cards. Data from these cards can be downloaded to a computer. Download data daily to ensure that the yield monitor is working properly and to protect against accidental data loss. Memory cards may store several megabytes of data. The card capacity is sometimes stated in hours of operation since data are typically stored on a periodic basis. One megabyte of memory can store 15–45 h of information for yield data collection intervals of 1–3 s.

11.4 Calibration and Yield Mapping Calibration

The types of calibration that are required by yield monitoring systems vary by monitor type. Regardless of the type of monitor, yields are not measured directly. Instead, measurements of force, displacement, or volume, speed of material flow, moisture content, harvester travel speed, and working width are combined to produce an estimate of crop yield. Crop yield is a derived, or calculated, value. Calibration is performed to ensure that sensor data and operator input are properly used by a monitor to produce a final output in units of harvest units per acre. In addition, force and displacement sensing grain flow meters must be calibrated to nullify, or cancel, the effects of machine vibration on yield readings. Most calibration processes involve comparing the weights of several loads estimated by a yield monitor with those measured on a separate set of scales. The actual weights are then entered into the yield monitor. Software within the yield monitor console then computes a set of calibration curves. To properly calibrate a yield monitor, the requirement is to harvest at different speeds, harvest widths, and a number of loads.

11.5 Limitations of Yield Monitors

The combine operation is dynamic, and the flow rate of material processed can vary depending on entering and exiting the crop. These varying flow rates can influence the results of the yield monitor data. Since the yield monitor measures the rate at which clean grain is entering the grain tank, time delays between the time grain enter the combine header, and the time it passes through the clean grain elevator can be significant. Combines also smooth abrupt changes in yield; hence, the yield monitor measures delayed averages of yield.

The phenomena of time delays and smoothing are most obvious when a combine enters or leaves the crop at the ends of a field. The combine, in the example above, has a delay of 15 s to process the entering crop and would travel 110 ft and harvest almost 0.04 of an acre before an accurate or stable yield is displayed on the yield monitor. Most yield mapping software compensates for equipment delays caused by the combine and corrects the yield data. The resulting yield map will not be perfect, but it will be very adequate for observing the magnitude and location of yield variability. Yield data combined with mapping software and positional data are capable of producing a colorful map.

11.6 Yield Mapping Concept

Yield mapping refers to the process of collecting georeferenced data on crop yield and characteristics, such as moisture content, while the crop is being harvested. Various methods, using a range of sensors, have been developed for mapping crop yields.

Yield maps are based on sets of instantaneous yield data points collected with the yield sensor compotes and a DGPS receiver (Fig. 11.5). The complete precision farming system consists of many elements and brings together a number of different technologies. The success of the precision farming system relies on the integration of all the different elements into a single system that can be operated at the farm level (Pfost et al. 1999). Many of the elements are required to be automatic. For example, the main task of the combine operator is to drive the combine to harvest the field and not to collect yield information used to generate yield maps. Therefore,

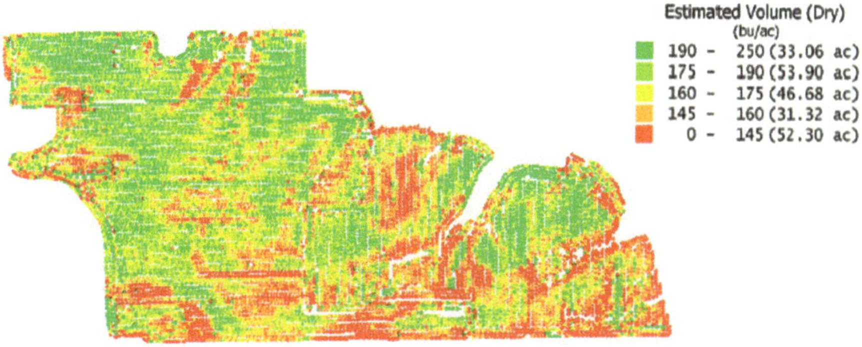

Fig. 11.5 Yield map

the yield mapping system employed to collect yield data must be automatic. Yield mapping is the process of continuously recording the grain flow through the combine harvester while at the same time integrating with actual harvest position in the field. The yield and position data are transferred to a computer where interpolation techniques are used to generate a yield map of the field. Once the data is collected, users need special computer software to generate yield maps. Most collected data are stored in a binary format. Binary files store digital data very efficiently, but this data must be converted to standard text files in order for most yield mapping software to process the data. The yield map provides information to the farmer detailing which are the better and poorer yielding parts of the field. The yield map does not provide information as the cause of the yield variation. The farmer is required to use a combination of local knowledge and other sources of data to interpret the maps to make crop management decisions to improve the profitability of the field.

11.7 Yield History Evaluation

Evaluating the temporal (year-to-year) variation of yield distribution within the field is an essential step in defining field areas with potentially high and low yields. Several approaches can be used to evaluate temporal effects on yield. One approach is to calculate the relative (normalized) yield for each point or grid cell. Normalized yield can be defined as the ratio of the actual yield to the field average.

When growing conditions in a field vary considerably, such as irrigated and dryland areas or different crops or varieties grown in different areas, normalization should be done separately for those areas, with the resulting relative yields recombined into one data file for the whole field. Figure 11.6 shows a relative yield history for a field with corn (soybean in the southern half in 2000) grown using furrow irrigation (until 2001) and center-pivot irrigation (in 2002).

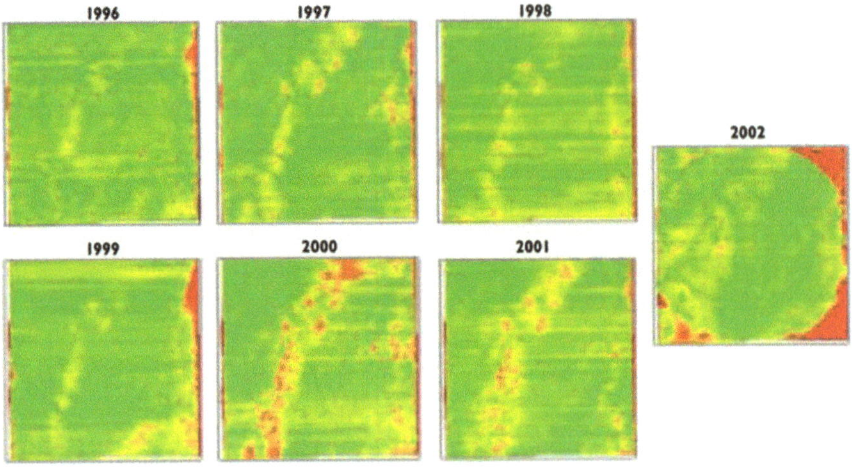

Fig. 11.6 Maps of relative yield of corn and soybean grown during a 7-year period (red indicates low-yielding areas and green indicates higher-than-average yields)

11.8 Potential Applications

Yield maps represent the output of crop production. On one hand this information can be used to investigate the existence of spatially variable yield-limiting factors. On the other hand, the yield history can be used to define spatially variable yield goals that may allow varying inputs according to expected field productivity.

The flowchart in Fig. 11.7 illustrates the process one might follow in deciding whether to invest in site-specific crop management, based on analysis of yield maps. If yield variability across the field cannot be explained by any spatially inconsistent field property, uniform management may be appropriate. Site-specific management becomes a promising strategy if yield patterns are consistent from year to year and can be correlated to one or more field properties (e.g., nutrient supply, topography, past management, etc.).

If the causes for yield variation are known and can be eliminated permanently, the entire area could be brought to similar growing conditions and managed uniformly thereafter. This concept was one of the earliest philosophies behind precision agriculture but is likely only feasible for certain field properties. For example, variable rate liming can be used to correct acidic areas in a field. In this case, the yield map is used only to investigate whether low soil pH is a yield-limiting factor, and the soil map is used to prescribe variable application rates. Another example would be localized deep soil tillage to alleviate compaction in selected field areas.

Most yield-limiting factors cannot be modified permanently through single measures because of economic or practical constraints. Consequently, site-specific crop management may be used to appropriately account for the existing spatial variability in attainable yield and/or soil properties.

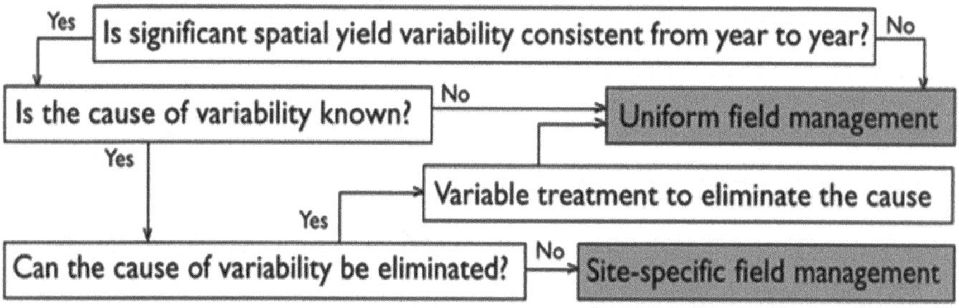

Fig. 11.7 Yield map decision-making flowchart

11.9 Conclusions

Yield maps are only as accurate as the data collected to produce them and only demonstrate that yield variability exists. Monitors must be correctly installed and periodically checked to provide accurate data. Yield map data should be used with soil test data, scouting notes, and other observations to learn why variability exists. The knowledge gained from site-specific crop management equips farmers to make better management decisions that have positive environmental benefits and that result in improved productivity and profitability.

References

Pfost, D., Casady, W., & Shannon, K. (1999). Global positioning system receivers. #6. In Clay et al. (Eds.), *Site-specific management guidelines*. Norcross: Potash & Phosphate Institute.

Thylen, L., & Murphy, D. P. (1996). The control of errors in momentary yield data from combine harvesters. *Journal of Agricultural Engineering Research, 64*(4), 271–278.

Wang, M. H. (1999). Field information collection and process technology. *Agriculture Mechanization, 7*, 22–24.

Chapter 12
Feasibility and Evaluation of Precision Farming

Abstract Precision farming is characterized by a number of sophisticated tools that assist in monitoring variation and managing inputs. Majority of farms in India are rain-fed, and with the change of weather patterns, availability of rainwater is very unpredictable. Hence the nonavailability of inputs and labor on time is the biggest stumbling block to increase productivity of farms and remuneration. PA can help in this matter. In India precision farming is still at a very nascent stage although it is very much in practice in developed countries. In India more than 57.8% of operational holdings have size less than 1 ha. However, in the major agricultural states of Punjab, Rajasthan, Haryana, and Gujarat, more than 20% of agricultural lands have operational holding size of more than 4 ha. But there are many constraints that are limiting adoption of site-specific crop management in Indian states like culture and perception of users, land fragmentation and small farm size, high initial costs, knowledge and technical gaps, lack of success stories, climatic aberrations, poor connectivity in rural areas, high energy demands, lack of local technical expertise, etc. This chapter focuses on feasibility of precision farming, the problems faced during adoption of precision farming techniques, and the potential in India. A section on feasibility in Jammu and Kashmir state has also been included.

In this chapter the economics and profitability of precision farming technologies are also analyzed. Rapid socioeconomic changes in some developing countries, including India, are creating new scopes for application of precision agriculture (PA).

Keywords Capital expenditures · Economics · Feasibility · Constraints · Profitability · Environmental benefits

12.1 Present Scenario of Agriculture in India

India is a global agricultural powerhouse. It is the world's largest producer of milk, pulses, and spices and has the world's largest cattle herd (buffaloes), as well as the largest area under wheat, rice, and cotton. India counts among the top two producers in the world of wheat, rice, sugarcane, cotton, milk, fruits, vegetables, and spices.

L. Ahmad, S. S. Mahdi, *Satellite Farming*,
https://doi.org/10.1007/978-3-030-03448-1_12

Table 12.1 India's rank in the production of various agricultural products

Agriculture production	India's rank	Next to
Wheat	Second	China
Rice	Second	China
Pulses	First	
Fruits and vegetables	Second	China
Sugarcane	Second	Brazil
Tea	Third	China, Turkey
Cotton (lint)	Second	China
Milk	First	

India's rank in the production of various agricultural products is given in Table 12.1. The country has some 195 M ha under cultivation of which some 63% are rain-fed (roughly 125 m ha) while 37% are irrigated (70 m ha). In addition, forests cover some 65 m ha of India's land. There are 118.9 million cultivators across the country or 24.6% of the total workforce of over 481 million (Census, 2011).

12.2 Problems in Indian Agriculture

India accounts for only about 2.4% of the world's geographical area and 4% of its water resources but has to support about 17% of the world's population and 15% of its livestock. Census 2011 says there are 118.9 million cultivators across the country or 24.6% of the total workforce of over 481 million. It has declined from nearly 50% in 1951 to 24% in 2011, which means the number of farmers has come down by half. About 70% of the rural households and 8% of urban households are still principally dependent on agriculture for employment. Since some three-quarters of the population live in rural areas, a majority of households thus depend principally on this sector. However, farming in India is non-remunerative, and 50% of farmers are in debt – the main reason for a large number of suicides because of poor availability of funds, farm inputs, poor support price structure for the produce, and almost no farm insurance. Agriculture contributes only about 14% to India's GDP.

Although rural India in aggregation has experienced its most dramatic period in the last 15 years, Indian agriculture productivity does not compare well with other countries, and the benefits have not permeated uniformly. Therefore, extreme poverty and distress are patently visible in rural India and among farmers. Added to this is the exodus of youth and their distance from agriculture, which have resulted in a peculiar contradiction – excessive population in the rural areas and yet shortages of labor for agriculture. The extreme weather conditions have also affected farmers.

In the existing agricultural scenario, India is characterized by small farms with around 80% of total holdings less than 2 ha (5 acres) and land mostly rain-fed with only 30% irrigated. Around 55% of the Indian population depends on farming, as against less than 4% in the United States and other advanced countries, because of heavy mechanization of agriculture. However, farming in India is non-remunerative and 50% of farmers are in debt – the main reason for a large number of suicides because of poor availability of funds, farm inputs, poor support price structure for the produce, and almost no farm insurance.

Also, without mechanization, farming is hard and back-breaking work. Farmers get more money in selling their land to builders, malls, and factories. This has put more pressure on farm land, thereby requiring technologies to increase the productivity so that shrinking farm land can feed billion plus people of India in the future.

India, though one of the biggest producers of agricultural products, has very low farm productivity, with the average of only 33% of the best farms world over. This needs to be increased so that farmers can get more profit from the same piece of land with less labor.

One of the biggest successes achieved by India postindependence has been its ability to move from facing food shortages and food imports to that of self-sufficiency and exports of food grains in a rather short period. Indian agriculture has also moved away from subsistence farming to intensive and technology-led cultivation. Still there are some serious issues faced by Indian agriculture:

1. Larger population dependent on agriculture – People are dependent on agriculture in India – ~55% of total employment across India, and it contributes only 14% to India's GDP.
2. Fragmented landholding and inadequate irrigation – More than 80% of farmers in India are marginal and small farmers. Inadequate irrigation makes it difficult for them to sustain farming.
3. Lack of institutional credit – Nearly 40% of all loans come from informal sources. Marginal landholding households suffer the most with only 15% of their credit from institutional sources such as the government, cooperatives, and banks – for households with land more than 10 ha the ratio is 79% (Sangwan 2015).
4. Output uncertainty – This is due to weather and other factors such as lack of market linkages, appropriate application of fertilizers, agrochemicals, etc.

12.3 Present Scenario of Precision Farming in India

The principle technology should be efficient, practical, cost-effective, and free from pollution. The sustainability factor should be looked at the ability of agricultural land to maintain acceptable levels of production over a long period of time, without degrading the environment. But the main drawback which the developing countries are facing especially India is that PF techniques are not properly adopted by the Indian farmers as they are financially weak and have small land size holdings. Still there is a hope that progressive Indian farmers, with guidance from the public and private sectors and agricultural associations, will adopt it in a limited scale for demonstrations as the technologies show potential for raising yields and economic returns with minimizing environmental degradation.

In India precision farming is still at a very nascent stage although it is very much in practice in developed countries. PA has been identified as one of the main thrust areas by the working groups (WGs) of India–US Knowledge Initiative on Agriculture (KIA). It is expected that PA research will be an important part of the recently launched ambitious agricultural research program, National Agricultural Innovation Project (NAIP), which will focus on innovations in agricultural technology with the announced budget of US$ 285 million. Space Application Center, ISRO, in collaboration with Central Potato Research Institute, Shimla, has initiated a study on exploring the role of remote sensing in precision farming. With the Space Application Center (ISRO), Ahmedabad has started experiments in the Central Potato Research Station farm at Jalandhar, Punjab, to study the role of remote sensing in mapping the variability with respect to space and time (Patil and Shanwad 2009).

M.S. Swaminathan Research Foundation Chennai, in collaboration with NABARD, adopted a village in Dindigul district of Tamil Nadu for variable rate input application, and the experiment was successful. Studies at London School of Economics, Harvard, Indian Institute of Management, Ahmedabad, and other premier research institutions show an increase of 80–600% increase in yields in different crops under Tamil Nadu precision farming project. Indian Agricultural Research Institute is doing precision farming on institutes' farms. Project Directorate for Cropping Systems Research, Modipuram and Meerut (Uttar Pradesh) in collaboration with the Central Institute of Agricultural Engineering (CIAE), Bhopal also initiated variable rate input application in different cropping systems. It's important here to mention about Tata Kisan Kendra (TKK). The concept of precision farming being implemented by the TKKs has the potential to catapult rural India from the traditional system to the world of IT and satellites. Polyhouse technology imbedded with precision farming principles has led to a new revolution in hill agriculture in Himachal Pradesh.

The PA technology is started to be developed and disseminated in a regionally differentiated manner through 17 Precision Farming Development Centers (PFDCs) located in different parts of India (Dugad and Sudhakar 2006). PFDCs are working for the popularization of PA and hi-tech applications to achieve increased production in addition to imparting training to a large number of farmers. But all these PFDCs mainly concentrate on precision irrigation water management. As an example of collaborative effort of private and government agencies, a new precision farming center has been established by MSSRF (M.S. Swaminathan Research Foundation – a nonprofit trust) at Kannivadi in Tamil Nadu with financial support from the National Bank for Agriculture and Rural Development (NABARD).

In the state of Jammu and Kashmir, precision farming has not been brought into practice yet. This presses upon the need of immediate implementation of IT-based farming in the state.

12.4 Present Scenario of Precision Farming in Jammu and Kashmir

Jammu Kashmir has varied topography, and agriculture remains the backbone of the economy of Jammu Kashmir as over 65% of the population depends on agriculture and allied sectors. Though the kind of diversity in agroclimatic conditions of Jammu and Kashmir is conducive for propagation of diversified farming system, the terrain at the same time is tough, and accessibility to a greater part of the region is poor. This causes a lot of hurdles in the movement of inputs, products, and technology. Over the years, the agriculturists and farmers have adopted several area-specific and time-specific cultivation practices to meet the requirement of their staple food crops. Rice, maize, wheat, pulses, fodder, oil seeds, potato, and barley are the main crops of the region.

However, there is a greater challenge; the state is facing low productivity of all agricultural crops, and there is massive deficit in its own production of food grains (40%), oil seeds (70%), and vegetables (30%). The net sown area of 7.52 lakh hectares (2004–2005) is 35% of the reported area as against national average of 46%.

Scientists world over have dug in their heads and worked hard for providing a solution to these challenges and have developed a meticulous system called precision agriculture/farming system. Since long, it has been recognized that the crops and soil are not uniform within a given field. The farmers have always responded to such variability, but such actions are inappropriate and less frequent.

Now modern agricultural management practices are changing from assuming homogenous fields to attempting to address field variability by dividing the field into smaller zones and managing these zones separately. Therefore, adoption of "precision agriculture" is the need of the hour which refers to the application of precise and proper amounts of inputs like water, fertilizers, etc. at the proper time to the crop for increasing its productivity and maximizing its yields.

All aspects of the environment soil, weather, vegetation, and water vary from place to place. All these factors determine crop growth and farming success. Precision farming allows precise mapping of all these factors. For small farms, precision agriculture as discussed earlier may include subsurface drip irrigation for precise water and fertilizer application and robots for no-till sowing, weed removal, harvesting, and other operations. These robots can run on renewable fuels and are compact. Hence, they do not need much maintenance or the arduous task of taking machines on lease.

As a result, the farmer would just have to identify better crops and invest more time into producing and experimenting widely distinct varieties of crops. Small farms are the principle spots for implementing large-scale precision agriculture. The government needs to provide subsidized loans to the farmers and guarantees farm insurance along with major funding in the R&D sector for implementing PF. Engineers have to design innovative drones and robots to ensure progress in precision agriculture. The industries and corporate should provide better job allowances so as to attract talented agriculture graduates.

12.5 Potential of Precision Farming in India with Special Reference to Jammu and Kashmir

Precision farming is thought of as a distant dream in developing countries, despite of its overwhelming success in developed countries. The biggest of all hurdles in adoption of the innovative farming techniques in India is the small field size. In India more than 57.8% of operational holdings have size less than 1 ha. However, in the major agricultural states of Punjab, Rajasthan, Haryana, and Gujarat, more than 20% of agricultural lands have operational holding size of more than 4 ha. So, there is a scope of implementing precision farming for major food grain crops at least in abovementioned states. Apart from the area consideration, other top factors impacting precision agriculture market in India and the world as a whole are growing adoption of smartphones, aggrandized use of cloud technology, booming population, and technological advancement.

Precision farming has a huge potential for improving the agricultural production in the developing world. Indian government has taken a bold initiative to double farmers' income by 2022. The union government aims to shift farm policies from production-oriented to income generating. In this regard, precision farming has emerged as a potential sustainable technology to achieve the goal. Precision farming complimented by right policy constitutes an important key to double the farmers' income by 2022. Precision farming is thought of as an effective tool to achieve double income as India has a latent huge resource of engineers, scientists, and agriculturists. These are lifeblood to develop various components of the technology as well as to deliver the technology.

Precision farming technologies can be used to double the area under irrigation in India without requiring extra water by improving water use efficiency. These technologies provide information to farmers on the optimal use of different fertilizers according to soil type, crop, and water usage. The rising concern for sustainability and efficiency necessitates harnessing of natural resource management technologies and improved practices. Modern machinery such as laser land levelers, self-propelled sprayers, precision seeders and planters, transplanters, etc. must be brought in use by the common farmers that will allow technically efficient farming and resource conservation. Adopting precision farming on a nationwide scale in the future will thus prove sustainable apart from being economical.

Globally, there are challenges arising from the anticipated global warming scenario and its possible impact on diverse agroecosystems, in terms of alterations in traditional crop belts mainly. This implies that the future specific crop production will be limited to certain areas only and that to non-traditional belts. Fourth Assessment Report of IPCC highlighted the contribution of agriculture in greenhouse gas production and figured it at 20%. From paddy fields, cattle rearing, chemical inputs to burning of farm waste, and other farm operations, agriculture is found to be one of the main contributors of greenhouse gas emission. As a common notion among farmers, tones of synthetic fertilizers are being used to increase production. This exorbitant and continuous use of synthetic fertilizers leads to excessive greenhouse gas (GHG) emission which is the main factor responsible for climate change due to human activities. In this regard precision farming proves a boon as it precisely uses fertilizers for optimum crop growth. This also protects environment from deteriorating effects of excessive chemical-based inputs. The use of precision farming on a large scale will thus be advantageous on environmental protection front as well.

The state of Jammu and Kashmir, located in the extreme western sector of the Himalayas, extends from 32°17′ to 37°50′ latitude and 72°40′ to 80°30′ longitude. Agriculture occupies an important place in the economy of state. Horticulture forms an important sector under the vast umbrella of agriculture in J&K state, especially Kashmir valley, with an annual turnover of fruits worth Rs. 3000 crores and giving employment to over 25 lakh people.

The key challenge confronting agriculture, in the state like other parts of country and world on a larger scale, is to produce more for a growing population and yet do so in a sustainable manner. Future projections reveal that apart from increased demand of food grains, there will be decrease in arable land, costly inputs, less water, poor soil health, and costly labor along with surging climate change. Thus, agriculture in J&K is at a crucial stage. Productivity boom should not mean harm to the natural resource bank. This can be achieved only with well-planned and well-executed developmental strategy, with conservation of natural resources and their rationale use on sustainable basis as the pole star. To achieve this goal, precision farming is going to help a lot in the future. Adoption of precision farming in the state will help realize the seemingly distant dream. Application of precision farming techniques will not only increase the productivity by providing optimum nutrients to the crop and precise management of pests and diseases but will also provide a safeguard to the natural wealth against rampant use of various chemical inputs. Less input application including irrigation water will reduce the cost of production while fetching a hefty amount due to quality produce. Water resources will not run out of water due to ever-increasing demand of irrigation.

J&K nestles in a fragile ecosystem and is highly vulnerable to ill impacts of climate change. The Fifth Assessment Report has emphasized the loss of ice from glacial range and projected that the current glacial extents are out of balance with climatic condition indicating that glaciers will continue to shrink in the future even without further temperature increase. Therefore, a comprehensive Climate Change Action Plan has been conceptualized by the state of J&K that will address these issues. Sustainable agriculture is one of the sectors identified as a part current of the Climate Change Action Plan. Among the list of key priorities of Sustainable Agriculture Mission, one features precision farming. It advocates the adoption of protected cultivation and precision farming under hi-tech green houses, poly green houses, and shade nets. Thus, in the future, it's expected and will be advantageous to replace conventional agricultural practices with precision farming technology.

The future direction of agriculture will depend upon the research community's ability to make the present technologies more and more user-friendly and manufacturers' will to provide these technologies at affordable prices to the farmers.

12.6 Economic Feasibility of Precision Farming in India on Agriculture Condition

Unlike some new technologies, there is no clear answer as to whether or not PA is economically beneficial. In Indian agriculture condition on the one hand, there are depletions of ecological foundations of the agroecosystems, as reflected in terms of increasing land degradation, depletion of water resources, and rising trends of floods, drought, and crop pests and diseases. On the other hand, there is imperative socioeconomic need to have enhanced productivity per units of land, water, and time.

At present, 3 ha of rain-fed areas produce cereal grain equivalent to that produced in 1 ha of irrigated out of 142 million ha. Net sown areas, 92 million ha, are under rain-fed agriculture in the county. From equity point of view, even the record agricultural production of more than 200 Mt. is unable to address food security issue. A close to 60 Mt. food grains in the storehouses of Food Corporation of India (FCI) is beyond the affordability and access to the poor and marginalized in many pockets of the country.

Globally, there are challenges arising from the globalization especially the impact of WTO regime on small and marginalized farmers.

Some other unforeseen challenges could be anticipated global warming scenario and its possible impact on diverse agroecosystems in terms of alterations in traditional crop belts, microlevel perturbations in hydrologic cycle, more uncertain crop–weather interactions, etc.

12.7 Scope of Precision Farming in Indian States

After green revolution, the Indian farmers and scientists understood that the soil is polluted due to excessive use of fertilizers and chemicals. Majority of the water is wasted leading to least resource management. In India it is the soil rather than the crop that is irrigated. As a result, 70% of the water is wasted. A modern technique which improves the utilization of resources optimally improves the quantity as well as the quality of crops, lowers the usage of fertilizers and pesticides, prevents soil degradation, and uses water more effectively and efficiently, thereby reducing the cost of production, and raising productivity has been introduced in India. This technique is none other than precision farming which includes the aid of modern and eco-friendly farming activities and technology, including satellite imagery and information technology. Precision farming enables the farmers to be precise in terms of time and space. "This innovation can go a long way in tackling many of our country's farm ills, including excessive use of water and other inputs, which has hurt soil quality apart from making farming unprofitable as a profession," says leading agricultural scientist M.S. Swaminathan, Chairman of the M.S. Swaminathan Research Foundation.

Jammu and Kashmir being the 11th largest state accounting for 3.20% of total area of the country with a geographical area of 103,187 sq. km situated at an altitude variation of 300–5400 meters above mean sea level constitutes about 67.5% of the Northwestern Himalayan region. Agriculture occupies an important place in the economy of the state. The share of agriculture and allied sectors in the gross state domestic product (at 1999–2000 prices) for the year 2010–2011 as per advanced 70% of the population in the state derives its livelihood directly or indirectly from the agriculture sector. In the state of Jammu and Kashmir, 58% of the area under agriculture is rain-fed, and the remaining 42% is irrigated. In spite of increase in food grain production from 4.52 lakh tons in 1950–1951 to 15.21 lakh tons, the demand is still deficit by 38%. The deficiency in cereals is nearly by 40%, 70% each in pulses and oil seeds and 30% in vegetables. The state still meets its demand by procuring large quantities from other states even after six decades of our independence. There is a greater challenge; the state is facing low productivity of all agricultural crops. The solution to this challenge lies in *precision farming*.

So, there is ample scope of precision farming in Jammu and Kashmir as well. The farmers can meet their desirable yields only by adoption of technology-based farming system which integrates the usage of GPS, GIS, remote sensing, LCC, VRT sensors, DSS tools, and other such systems. The problem in adopting such a technique of crop production lies in its high money demand, but the farmer has to demonstrate that precision farming is cost-effective, reducing the wastage of resources and other inputs. They need to be guided in the usage of such high-tech equipment which are site specific and help the farmers to obtain more revenue. The farmers need to spend their money in a wise manner on their fields to achieve high benefits and profits by the application of precision farming. Although such a system is not found in state yet, the farmers have to realize the need for an important farming system to increase their productivity and achieve sustainability in their production system through application of precision farming.

Precision farming is one of the recent technologies, which is implemented in the South Indian state of Tamil Nadu, with a main focus on increasing the productivity of fruits, vegetables, and flowers. Among the vegetables, tomato is an important crop both in terms of its area in the State and its importance in the food preparations of the people of India. It is important to assess the technical efficiency of precision and non-precision (conventional) farming and also to compare the profitability of the tomato cultivation, so as to recommend the precision farming technology for wider adoption at national level. Our study reveals that the net returns realized per ha of tomato under precision farming technology (Rs. 240,455) were found to be 300% higher than that of non-precision technology (Rs. 82,600). The yield gap analysis revealed that about 40% of the difference in yield between precision and non-precision farming is due to the use of advanced technology.

12.8 Constraints in the Adoption of Precision Agriculture in India

In the present existing situation, the potential of precision agriculture in India is limited by the lack of appropriate measurement data management tools that must be developed and validated to support both research and production. The limitation in data quality/availability has become a major obstacle in the demonstration and adoption of the precision technologies. The review of the past studies has shown that the lack of finance and credit facilities were the most important obstacle for non-adoption of precision farming, as obtaining credit was a difficult process because farmers could not produce collateral security. The following are the major constraints in the adoption of precision agriculture in India.

12.8.1 Culture and Perception of Users

Farmers do not make changes easily. Especially, the traditional farmers are reluctant to adopt new practices. The basic reason for this is that the farmers have survived their own way of doing. They do not permit any change in the long-established methods that they have been practicing. These practices have fetched satisfying results. Hence, farmers are very comfortable with their cultural practices. Therefore, there are some societal constraints related to adoption of innovative agriculture.

12.8.2 Land Fragmentation and Small Farm Size

In India farming is mostly subsistent and landholding size is small. Land fragmentation is believed to be the chief obstacle in the extensive adoption of precision agriculture in India. This fragmentation includes problems like small size and irregular shape of farms. Since the farm size is significantly smaller, the challenges are bigger. This can discourage the uptake precision farming. Searching for the right type of precision farming technologies for small farmers is a big challenge.

12.8.3 High Initial Costs

The advanced precision farming technologies require large upfront investments. This is because the precision farming tools and technologies are generally high-priced. On the contrary, majority of the farmers in India are poor. These expenses of precision farming instruments and services are difficult to be met by the common Indian farmers, unless some start-ups are launched to help farmers get into precision farming.

12.8.4 *Knowledge and Technical Gaps*

Though precision agriculture seems to have a great scope in a country like India, the basics of it are not yet known to the common Indian farmers. Most of the farmers are still unaware of the use of computers in agriculture and unheard of the various technologies of precision agriculture like GPS (global positioning system), GIS (geographical information system), RS (remote sensing), etc. There is not only a lack of awareness but also a lack of understanding. This gap, which exists in the knowledge of Indian farmers, restricts adoption of precision farming in India.

12.8.5 *Lack of Success Stories*

So as to get some motivation for the adoption of precision farming, a farmer would want to look for some motivational cases, like some successful examples. But unfortunately, the benefits of precision farming are not immediately visible, at the beginning of adoption. These benefits show up in the long run. This is why the farmers are rather reluctant to implement precision technologies on their farms and regard the use of it to be doubtful and uncertain. One more problem is the unwillingness of the producers to place the study designs on their farms without assurance of a profit margin.

12.8.6 *Climatic Aberrations*

At many places of India, the local agroclimatic conditions make the adoption of precision farming technologies unfeasible. In particular, the Himalayan region which includes the state of Jammu and Kashmir and others, the terrain and is rough; resulting in poor access to majority of the areas. This causes a lot of difficulty in the movement of inputs and technology. Besides this, various ecological constraints operate that abate the ability of precision agriculture techniques. Even the primary weather factors can cause GPS errors like loss of signal reception and ranging errors.

12.8.7 Poor Connectivity in Rural Areas

In India, most of the farming areas are located in remote areas that face poor network performance and bandwidth speeds. In a majority of Indian villages, a reliable Internet connection is not available. This even discourages the application of precision agriculture in such places. In the hilly terrain of some parts of India, reception of GPS is also poor.

12.8.8 High Energy Demands

Precision farming is "greener." It is well known that implementation of precision farming techniques provide various advantages to the environment and help in its sustenance. However, this type of farming is high energy demanding. The high-tech tools used in the precision farming operations are energy consumptive. In a developing country like India, where electricity supply in most areas is inadequate for household needs, energy requirements of precision agriculture are difficult to meet.

12.8.9 Lack of Local Technical Expertise

In order to apply precision farming technologies, different types of software are being used. Such software collect data, generate maps (e.g., yield maps, soil maps, and variable rate application maps), overlay these maps, and generate geostatistical results. Besides these "soft" components, precision farming includes the use of hi-tech "hard" components. Some of these are still very complex and complicated for Indian farmers. There is a lack of precision know-how in India. This lack of technical skill and unawareness related to operation of tools is a constraint for the adoption of precision farming in India.

12.8.10 Poor Assistance from Government

Adoption of precision farming by Indian farmers is unlikely if the government does not step in. Assistance to the farmers, especially to small and marginal farmers, is necessary. Though some organizations like ISRO, M. S. Swaminathan Research Foundation, Chennai and IARI, and New Delhi have taken a step in this direction, these efforts are still insufficient.

12.8.11 Infrastructure and Institutional Constraints

Precision agriculture is still in its infancy in India. Unlike the developed countries of the world, Indian agriculture suffers the absence of strong working groups. There is a need of some discrete initiative to build up a strong infrastructure for precision farming. Development of specific centers and scientific data banks is a precondition for precision farming. Collaborative efforts from private and government agencies are needed which unfortunately are not adequate in India.

12.8.12 Other Constraints Include the Following:

- Heterogeneity of cropping systems
- Operational constraints
- Market imperfections
- Uncertainty in getting various inputs
- Absence of long-standing and uniform agricultural policy

The adoption of precision agriculture needs combined efforts on behalf of scientists, farmers, and the government. The following methodology could be adopted in order to operationalize precision farming in the country:

1. Creation of multidisciplinary teams involving agricultural scientists in various fields, engineers, manufacturers, and economists to study the overall scope of precision agriculture.
2. Formation of farmer's cooperatives since many of the precision agriculture tools are costly (GIS, GPS, RS, etc.).
3. Government legislation restraining farmers using indiscriminate farm inputs and thereby causing ecological/environmental imbalance would induce the farmer to go for alternative approach.
4. Pilot study should be conducted on farmer's field to show the results of precision agriculture implementation.
5. Creating awareness among farmers about consequences of applying imbalanced doses of farm inputs like irrigation, fertilizers, insecticides, and pesticides.
6. Some farm holdings are fragmented, and even minimal mechanization is not possible.
7. The soil and other analytical facilities are also a limiting factor.
8. Culture and perceptions of the users – high-tech cultivation practices are not accepted by farmers in developing countries because the farmers are largely resource poor.
9. Most of the inputs, namely, water-soluble fertilizers, hybrid seeds, annual fruit seeds, and seedlings are being used at high costs, and only a few farmers can afford the costs involved.

10. The existing post harvesting facilities and market chains are at stake and not in a position to store or transport the produce.
11. High cost of sophisticated machines/equipment also acts as a limiting factor.
12. Small farm size.
13. Lack of success stories.
14. Heterogeneity of cropping systems and market imperfections.
15. Land ownership, infrastructure, and institutional constraints.
16. Lack of local technical expertise.
17. Knowledge and technical gaps.
18. Data availability, quality, and costs.

As in the United States and Europe, the development and adoption of precision agriculture in India is a slow process. The small size of farms and fields in most of Indian agriculture limits economic gains from currently available precision farming technology, while the population density and public concerns for the environment, food safety, and animal welfare mean that those potential benefits of precision agriculture are being given more attention.

12.9 Economic and Environmental Evaluation of Precision Farming

In the implications of dramatic shifts for economic development, urbanization and energy consumption in some developing countries are immense. High-tech nature of traditional PA technologies developed in advanced countries created a real challenge for engineers to search suitable PA technologies for developing countries. It is expected that application of balanced soft and hard PA technologies based on the need of specific socioeconomic condition of a country will make PA suitable for developing countries also. The impact of adoption of precision on agriculture on environment has also been discussed.

One of the chief reasons for low adoption of PA is the reluctance of farmers to invest many thousands of rupees in PA without knowing if the technology will return a profit. A range of factors affect the investment value of PA including the current farm gross margin, cost of PA equipment, the area and number of years over which the equipment is used, and the rate at which benefits from adoption start to occur (Jennings 2005).

12.10 Capital Expenditures in Precision Agriculture

PA investments include purchases of equipment, installation charges, and the time and effort spent learning how to use and maintain the technologies. These costs are usually not recoverable if use is discontinued (Pindyck 1988). Compared to some other capital items with active resale markets, like land and tractors, PA equipment is highly specialized with limited resale potential and is usually a sunk cost. There might even be costs to uninstall or discontinue the use of precision equipment. Outsourcing to a custom service provider is another option, also with costs. These factors increase the financial risks of PA adoption. To offset these costs, farmers require a higher expected return to adopt PA technologies (Schimmelpfennig and Ebel 2016).

Precision agriculture technologies can affect farm finances through operating costs, overhead costs, and changes in yield. PA technologies could reduce operating costs and raise allocated overhead (by substituting capital and labor for operating inputs), or they could do the reverse. PA adoption could raise costs through increased use of inputs (if a need is indicated by mapping) but also raise yields. Since the impact of the technologies on operating costs alone would ignore this fundamental trade-off between capital and labor, net returns are considered first, beginning with the relative use of hired and unpaid labor.

Adoption of PA technology naturally leads to greater expenditures on machinery and equipment as these technologies are capital intensive. Machinery also has a higher expense base (compared to labor costs) and more potential to influence overhead costs.

12.11 Farm Size and Technology Adoption

Farm size is an important determinant of the adoption of technology (Feder et al. 1985). Farm size can affect and in turn be affected by the other factors influencing adoption (Lavison 2013). Some technologies are termed as scale-dependent because of the great importance of farm size in their adoption (Bonabana- Wabbi 2002). Because the fixed cost per unit of land decreases in scale, there is a threshold of farm size below which it is less profitable to adopt a technology (Feder and O'Mara 1981). The fixed costs will be high if the subcomponents are forever changing to adapt to the changing environment or new components being introduced where decision-makers have low levels of human capital (Schultz 1975), as is often the case in developing economies. This critical farm size decreases over time as uncertainty is reduced because of learning and the dissemination of information from early adopters.

Many studies have reported a positive relation between farm size and adoption of agricultural technology. The fixed transaction costs decrease with an increase in farm size, which explains why the adoption of new divisible technology might begin with larger farmers. Second, because larger farmers have a greater ability to raise capital, to bear the cost of the innovations, and to bear the risk of failure, they are likely to be less risk-averse compared to relatively small farmers and are more likely to make risky investments compared to smaller farmers.

Some studies have shown a negative influence of farm size on adoption of new agricultural technology. Small farm size may provide an incentive to adopt a technology especially in the case of an input-intensive innovation such as a labor-intensive or land-saving technology. Farmers with small land may adopt land-saving technologies such as greenhouse technology and zero grazing among others as an alternative to increased agricultural production (Harper et al. 1990).

A key determinant of the adoption of a new technology is the net gain to the farmer from adoption, inclusive of all costs of using the new technology (Foster and Rosenzweig 1995). The cost of adopting agricultural technology has been found to be a constraint to technology adoption.

12.12 Profitability

For a new practice to be widely adopted in production agriculture, the practice must yield an economic profit except for regulatory requirements. Variable rate application requires additional costs associated with soil sampling, data analysis, and new equipment. Economic benefits from variable rate application can only be derived from increased yields and/or savings in reduced inputs or a combination of both. If the initial investment for equipment IS high, actual economic returns of variable rate application will be low or even negative for the first few years. Some costs associated with soil sampling for a field can be accurately determined, whereas other costs for new equipment and data analysis are difficult to estimate tar each field. Nevertheless, it is certain that these costs will go down if the same equipment and data analysis software are used for more fields over more years.

Many studies have shown that variable rate applications of fertilizers are superior to uniform rate application in terms of economic benefits, but evidence of profitability has been mixed based on nine published field research studies (Swinton and Lowenherg-DeBoer 1998). Numerous recent studies present more evidence of profitability for variable rate application of fertilizers, pesticides, and seeds.

The measurable benefits of precision agriculture on the farm level are often separated in three different aspects: costs reduction, yield increase, and quality improvement. However, from a societal point of view, the benefits of precision agriculture serve more abstract and less measurable goals such as:

- **Increase of productivity:** The productivity of a farm (or a farmer), in this case not expressed as tons/ha but rather as the gross margin between costs and financial benefits per unit of labor, unit of equipment or unit of inputs.
- **Reducing risks:** Precision agriculture practices (including the application of sensors and precision equipment) help farmers to avoid risks or to respond adequately to them. Most prominent are the weather-related risks (both drought and water excess) and disease-related risks.
- **Improving transparency:** The so-called license to operate for farmers requires more and more proof of compliance to regulations or (quality) claims. Precision agriculture technologies help farmers to automatically document their activities

and as such reduce the administrative burden to get or keep their license to operate.

And behind those farm-related benefits, there are also societal benefits that align with these benefits, such as related to climate mitigation, food security, higher sustainability, and lower environmental footprints. These benefits are difficult to measure but are however closely linked with the abovementioned abstract farm goals: for instance, the increase of productivity expressed as "more crop per drop" (kg crop/m^3 water) has a positive effect on farm income (reduced cost per unit), hence on productivity, but also on risk reduction and therefore on climate mitigation and environmental footprint.

12.13 Environmental Benefits

Precision agriculture technologies are known to be environmentally friendly, since only the needed amount of agricultural chemicals (pesticides, fertilizers, etc.) is applied to areas as needed (site-specific treatment). Thus, the technologies help maintain the quality of the environment and the sustainability. One of the goals of precision agriculture is to optimize crop inputs so that the environmental impact could be minimized. By optimizing the input, growers can increase the efficiency of chemical applications, reduce time and labor, and thereby increase profit. Variable rate technology or precision application of agricultural chemicals can help increase environmental benefits as well as reduce waste.

The National Research Council (1997) reported that the environmental benefits of precision agriculture may not be a primary reason for the technology adoption, since growers are depending on economic savings or more profit. Precise calibration of input materials will increase plant uptake of those inputs, and precise fertilizer application will not leave any excess amount in the soil, thus reducing runoff and contamination to the groundwater (Thrikawala et al., 1999). The council recommended that synergy between biotechnology and variable rate technology could increase environmental benefits. It described potential environmental problems, including adverse effects from variable rate application to steeper slopes, more fertilization to potentially incorrectly interpreted high yielding areas, and improper use of environmental data by the regulating agencies.

Precision conservation can account for variability in topography, length, slope, hydrology, soil cover parameters, and other chemical and physical properties to implement best conservation and management practices. These procedures can be used to reduce off-site transport of nutrients and sediments from fields to surrounding areas and help manage field off-site areas, buffer areas, water channels, and other areas of the watershed. Not only will this reduce the further transport of sediments but also will contribute to minimizing agrochemicals entry into water bodies (Berry et al. 2003).

12.14 Conclusions

Precision agriculture contributes to sustainable intensification by enabling farmers to increase their yields with fewer inputs than other application methods such as broadcasting (scattering over a large area) fertilizers or seed, for example. This can also improve soil quality and moisture while minimizing the environmental impact that excessive input use can cause. Further, the targeted application of inputs can help farmers to be more competitive by lowering production costs. Precision agriculture involves measures such the targeted use of inputs through microdosing, soil testing, and proper seed spacing as opposed to broadcasting. This also reduces the impact on environment.

References

Berry, J. K., Detgado, J. A., Khosla, R., & Pierce, F. J. (2003). Precision conservation for environmental sustainability. *Journal of Soil and Water Conservation, 58*(6), 332–339.

Bonabana-Wabbi J. (2002). *Assessing factors affecting adoption of agricultural technologies: The case of Integrated Pest Management (IPM) in Kumi District.* Msc. Thesis Eastern Uganda.

Dugad, S. V., & Sudhakar, M. S. (2006). Application of information technology in irrigated agriculture. In *Proceedings of 19th national convention of agricultural engineers on role of information technology in high-tech agriculture and horticulture,* Bangalore, India (pp. 197–202).

Feder, G., & O'Mara, G. T. (1981). Farm size and the diffusion of green revolution technology. *Economic Development and Cultural Change, 30*(1), 59–76.

Feder, G., Just, R. E., & Zilberman, D. (1985). Adoption of agricultural innovations in developing countries: A survey. *Economic Development and Cultural Change, 33*(2), 255–298.

Foster, A. D., & Rosenzweig, M. R. (1995). Learning by doing and learning from others: Human capital and technical change in agriculture. *Journal of Political Economy, 103*(6), 1176–1209.

Harper, J. K., Rister, M. E., Mjelde, J. W., Drees, B. M., & Way, M. O. (1990). Factors influencing the adoption of insect management technology. *American Journal of Agricultural Economics, 72*(4), 997–1005.

Jennings, G. (2005). Calculating the costs of precision ag. *Grain Business* (pp. 16–17).

Lavison, R. (2013). *Factors influencing the adoption of organic fertilizers in vegetable production in Accra.* Msc Thesis, Accra Ghana.

National Research Council. (1997). *Precision agriculture in the 21st century: Geospatial and information technologies in crop management.* Washington, DC: Nationl Academy Press.

Patil, V. C., & Shanwad, U. K. (2009, December). Relevance of precision farming to Indian agriculture. In *Second national conference on agro-informatics and precision farming.*

Pindyck, R. S. (1988). Irreversible investment, capacity choice, and the value of the firm. *American Economic Review, 78*(5), 969–985.

Sangwan, S. S. (2015). More credit for agricultural households? NSSO's 70th Round on indebtedness. *Economic and Political Weekly, 50*(16), 23–24.

Schimmelpfennig, D., & Ebel, R. (2016). Sequential adoption and cost savings from precision agriculture. *Journal of Agricultural and Resource Economics, 41*(1), 97–115.

Schultz, T. W. (1975). The value of the ability to deal with disequilibria. *Journal of Economic Literature, 13*(3), 827–846.

Swinton, S. M., & Lowenberg-DeBoer, J. (1998). Evaluating the profitability of site-specific farming. *Journal of Production Agriculture, 11*(4), 439–446.

Thrikawala, S., Weersink, A., Fox, G., & Kachanoski, G. (1999). Economic feasibility of variable-rate technology for nitrogen on corn. *American Journal of Agricultural Economics, 81*(4), 914–927.

Chapter 13
Decision Support System for Precision Farming

Abstract As agriculture becomes more intensive, the demand for a higher level of control of the environment in which the plants grow increases. This control ranges from better strategies of soil management to "closed" environments, where most, if not all, atmospheric and soil variables can be adjusted. Based on this premise, plant growth and development models should be elaborated to supply a basis for planning and managing crop production. Crop modeling can also be useful as a means to help the scientist define research priorities. Using a model to estimate the importance and the effect of certain parameters, a researcher can observe which factors should be more studied in future research, thus increasing the understanding of the system. The model has also the potential of helping to understand the basic interactions in the soil–plant–atmosphere system. In this chapter, the reader can find a description of different crop simulation models, their types, and application in precision agriculture.

Keywords Model · Simulation · Production level crop model · DSSAT · InfoCrop

13.1 Model

A model is a schematic representation of the conception of a system or an act of mimicry or a set of equations which represent the behavior of a system, with the purpose of aiding, understanding, and improving the performance of the system.

13.2 Complexity of Agricultural Systems

Agricultural systems are characterized by having many organizational levels. From the individual components within a single plant, through constituent plants, to farms or a whole agricultural region or nation lies a whole range of agricultural systems. Since the core of agriculture is concerned with plants, the level that is of main interest to the agricultural modeler is the plant. Reactions and interactions at the level of

L. Ahmad, S. S. Mahdi, *Satellite Farming*,
https://doi.org/10.1007/978-3-030-03448-1_13

tissues and organs are combined to form a picture of the plant that is then extrapolated to the crop and their output.

13.3 Crop Modeling

Modeling is the use of equations or sets of equations to represent the behavior of a system. In effect crop models are computer programs that mimic the growth and development of crops (Oteng-Darko et al. 2013b). The model simulates or imitates the behavior of a real crop by predicting the growth of its components, such as leaves, roots, stems, and grains. Thus, a crop growth simulation model not only predicts the final state of crop production or harvestable yield but also contains quantitative information about major processes involved in the growth and development of the crop. Reactions and interactions at the level of tissues and organs are combined to form a picture of the crop's growth processes.

13.4 Types of Models

Depending upon the purpose for which it is designed, the models are classified into different groups or types. A few of them are as follows (Oteng-Darko et al. 2013a):

1. *Empirical Models*

These are direct descriptions of observed data and are generally expressed as regression equations (with one or a few factors) and are used to estimate the final yield. This approach is primarily one of examining the data, deciding on an equation or set of equations and fitting them to data. These models give no information on the mechanisms that give rise to the response. Examples of such models include those used for such experiments as the response of crop yield to fertilizer application, the relationship between leaf area and leaf size in a given plant species, and the relationship between stalk height alone or coupled with stalk number and/or diameter and final yield.

2. *Mechanistic Models*

A mechanistic model is one that describes the behavior of the system in terms of lower-level attributes. Hence, there is some mechanism, understanding, or explanation at the lower levels (e.g., cell division). These models have the ability to mimic relevant physical, chemical, or biological processes and to describe how and why a particular response occurs. The modeler usually starts with some empirism, and as knowledge is gained, additional parameters and variables are introduced to explain crop yield. The system is therefore broken down into components and assigned processes.

3. *Static and Dynamic Models*

A static model is one that does not contain time as a variable even if the end products of cropping systems are accumulated over time. In contrast dynamic models explicitly incorporate time as a variable, and most dynamic models are first expressed as differential equations.

4. *Deterministic Models*

A deterministic model is one that makes definite predictions for quantities (e.g., crop yield or rainfall) without any associated probability distribution, variance, or random element. However, variations due to inaccuracies in recorded data and to heterogeneity in the material being dealt with are inherent to biological and agricultural systems (Brockington 1979).

In certain cases, deterministic models may be adequate despite these inherent variations, but in others they might prove to be unsatisfactory, e.g., in rainfall prediction. The greater the uncertainties in the system, the more inadequate deterministic models become.

5. *Stochastic Models*

When variation and uncertainty reach a high level, it becomes advisable to develop a stochastic model that gives an expected mean value as well as the associated variance. However, stochastic models tend to be technically difficult to handle and can quickly become complex. Hence, it is advisable to attempt to solve the problem with a deterministic approach initially and to attempt the stochastic approach only if the results are not adequate and satisfactory.

6. *Simulation Models*

These form a group of models that is designed for the purpose of imitating the behavior of a system. Since they are designed to mimic the system at short time intervals (daily time-step), the aspect of variability related to daily change in weather and soil conditions is integrated. The short simulation time-step demands that a large amount of input data (climate parameters, soil characteristics, and crop parameters) be available for the model to run. These models usually offer the possibility of specifying management options, and they can be used to investigate a wide range of management strategies at low costs.

7. *Optimizing Models*

These models have the specific objective of devising the best option in terms of management inputs for practical operation of the system. For deriving solutions, they use decision rules that are consistent with some optimizing algorithm. This forces some rigidity into their structure resulting in restrictions in representing stochastic and dynamic aspects of agricultural systems.

8. *Weather Data for Modeling*

In crop modeling the use of meteorological data has assumed a paramount importance. There is a need for high precision and accuracy of the data. The data obtained from surface observatories has proved to be excellent. But the automated stations are yet to gain popularity in the underdeveloped and developing countries. As of now, solar radiation, temperature, and precipitation are used as inputs in DSSAT.

In crop modeling weather is used as an input. The available data ranges from 1 s to 1 month at different sites where crop modeling work in the world is going on. Different curve fitting techniques, interpolation, extrapolation functions, etc., are being followed to use weather data in the model operation. Agrometeorological variables are especially subject to variations in space. It is reported that, as of now, anything beyond daily data proved unworthy as they are either overestimating or underestimating the yield in simulation.

Role of Weather in Decision-Making

The principal effects of weather on crop growth and development are well understood and are predictable. Crop simulation models can predict responses to large variations in weather. At every point of application, weather data are the most important input. The main goal of most applications of crop models is to predict commercial output (grain yield, fruits, root, biomass for fodder, etc.).

Crop simulation models are used in the United States and in Europe by farmers, private agencies, and policy-makers to a greater extent for decision-making. Under Indian and African climatic conditions, these applications have an excellent role to play. The reasons being the dependence on monsoon rains for all agricultural operations in India and the frequent dry spells and scanty rainfall in crop-growing areas in Africa. Once the arrival of monsoon is delayed, the policy-makers and agricultural scientists in India are under tremendous pressure. They need to go for contingency plans. These models enable to evaluate alternative management strategies, quickly, effectively, and at no/low cost. To account for the interaction of the management scenarios with weather conditions and the risk associated with unpredictable weather, the simulations are conducted for at least 20–30 different weather seasons or weather years. If available, the historical weather data, and if not weather generators, are used presently. The assumption is that these historical data will represent the variability of the weather conditions in future. Weather also plays a key role as input for long-term crop rotation and crop sequencing simulations.

Some crop models reported in recent literature are (Kumar and Chaeturvedi 2009):

Software	Full form	Details
REALSOY	–	Soya bean
MODVEX	MOdel Development and Validation EXpert	Model development and validation system
IRRIGATE	–	Irrigation scheduling model
COTTAM	–	Cotton
APSIM	Agricultural Production Systems sIMulator	Modeling framework for a range of crops
GWM	General Weed management Model	General weed model in row crops
MPTGro		*Acacia* spp. and *Leucaena* spp.
WOFOST	WOrld FOod STudies	Wheat and maize, water, and nutrient
WAVE	Water and Agrochemicals in the soil, crop and Vadose Environment	Water and agrochemicals

Software	Full form	Details
SUCROS	Simple and Universal Crop gROwth Simulator	Crop models
CERES-Rice	Crop Environment REsource Synthesis	Rice, water
EPIC	Erosion Productivity Impact Calculator	predicts the impact of erosion on soil productivity
CERES	Crop Environment REsource Synthesis	Series of crop simulation models
DSSAT	Decision Support System for Agrotechnology Transfer	Framework of crop simulation models including modules of CERES, CROPGRO, and CROPSIM

13.5 Crop Production Levels

Based on growth-limiting factors, viz., water and nutrients, four levels of crop production have been identified.

Production Level 1 This is an ideal situation in which the crop does not suffer from any shortage of water or nutrients in the entire growing season. Its growth and development depend only on the weather conditions and crop characteristics. Figure 13.1 is a flow diagram showing the processes that are to be considered at this level (Singh 1994).

Production Level 2 The growth of a crop is limited by shortage of water for at least some part of the growing season. This situation frequently occurs in semiarid regions and also in areas where the rainfall is inadequate and/or poorly distributed.

Production Level 3 Nitrogen shortage during the growing season limits the crop growth at this production level. Water shortage or adverse weather conditions occur in the remaining part of the season. This condition is very frequent in the agricultural systems all over the world.

Production Level 4 Crop growth is restricted by low levels of phosphorus and other mineral nutrients in the soil during the growing season.

Growth-reducing factors such as diseases, insect pests, and weeds can occur at each of these production levels giving them an extra dimension. Simulating crop growth, therefore, implies that for an explanatory crop growth model, researchers from diverse disciplines must work in close, self-imposed coordination to achieve a common goal.

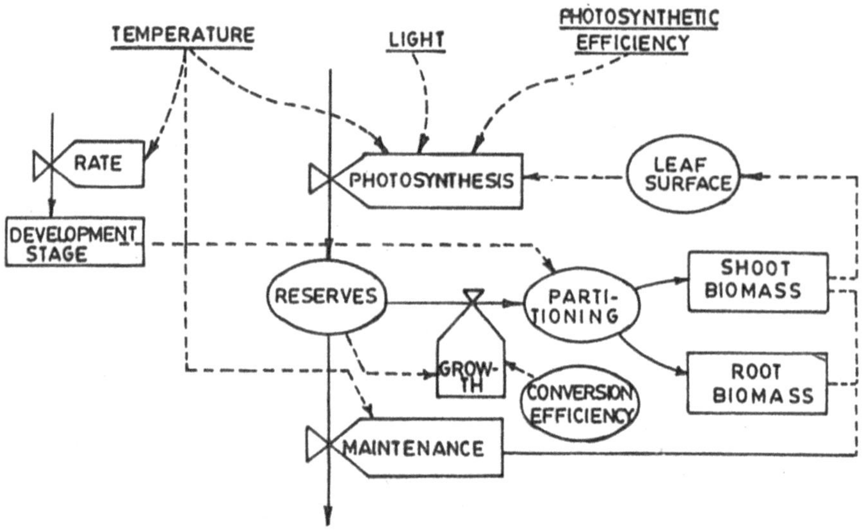

Fig. 13.1 Flow diagram showing the processes involved at crop production level 1

13.6 Applications and Uses of Crop Growth Models

Simulation modeling is increasingly being applied in research, teaching, farm and resource management, and policy analysis and production forecasts. They can be applied to, namely, research, crop system management, and policy analysis.

1. *Environmental Characterization*

Crop models together with GIS can greatly facilitate demarcation of homologous zones at mega-, macro-, meso-, as well as microlevel depending upon the availability of data and objectives. These tools have been used to determine potential and attainable yields for a given level of inputs for various crops. Potential yield of cultivars varies with season/year and location. Estimates of such yields for different varieties can establish a reference point for site quality and remove the confounding effects associated with large climatic variation.

2. *Optimizing Crop Management*

Once potential yields have been quantified, these can be converted to attainable yields to determine magnitudes of yield gap. Crop growth modeling can be used in matching agro-technology with the farmers' resources and in analyzing the precise reasons for yield gap.

3. *Pest and Disease Management*

At a regional level, GIS, requisite environmental data coupled with epidemic simulation models, further provide geographic delineation of disease and insect pest risk zones. These zones can also help us in making strategic decisions on deployment strategies for varieties and to determine how long host plant resistance would be expected to last. Physiologically based simulation models can be applied to understand the damage mechanisms and analyze their effects on crop growth and yield of rice.

4. *Impact of Climate Change*

Crop models are being used to estimate the impact of increased carbon dioxide and temperature on crop production (Matthews et al. 1997). It was assumed that the trends in potential yields would also be shown in actual yield. These models can utilize the input from Global Circulation Models (GCM) to quantify the impact of climate changes.

5. *Yield Forecasting*

In this approach, the model is run using actual weather data during the cropping season for the geographical region of interest. Weather data for typical years are used to continue simulations until harvest. A schematic diagram of the various components of such a system would be integrated as shown in Fig. 13.2 (Singh 1994).

6. *Optimal Sowing Dates for Hybrid Seed Production*

A major constraint to utilizing hybrid seed technology is the need to buy fresh seeds every planting season and the high cost of seed production. This limited seed production is largely due to poor outcrossing and asynchronous flowering of male and female parents. Environmental factors such as temperature, humidity, and wind speed at the time of pollination and fertilization play a great role in regulating outcrossing percentage. To synchronize these at several locations to determine the optimal planting dates, crop simulation modeling is a more efficient tool in making decisions about the planting calendar of parents in different environments (Xu 1995).

7. *Increased Efficiency of Multi-environment Evaluation*

Development and release of a variety is a complex process that may extend over a period of 10–15 years. Once the breeding lines have become homozygous, they

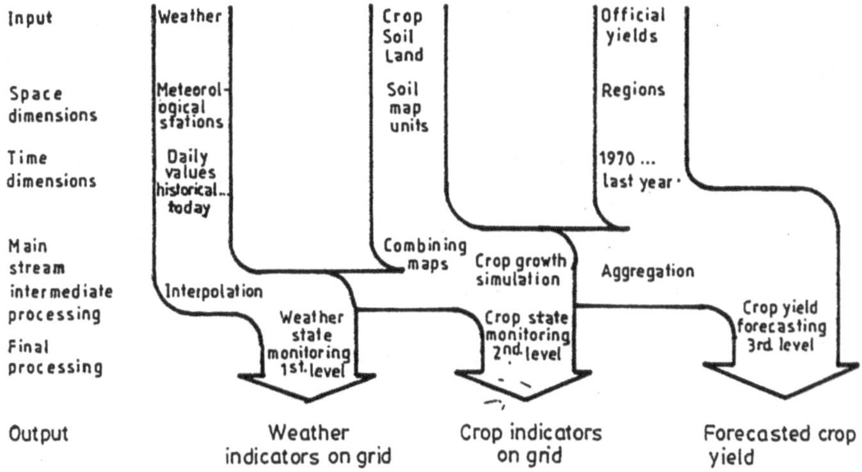

Fig. 13.2 A crop growth monitoring and yield forecasting system

are bulked and then tested in observational, replicated, and multilocation yield trials. These multi-site trials are expensive and need several crop seasons/locations to understand genotype by environment (G x E) interactions.

Since the systems approach integrates different components of agroecosystems, it can play a great role in increasing the efficiency of plant breeding process, in particular multi-environment testing of genotypes. Crop models together with GIS can facilitate biological characterization of the physical environment (geography, soil, climate, etc.) and thus define key environmental domains for which improved varieties are to be developed. Alternatively, the same methodology can be used to determine the adaptation domains of genotypes. A modeling approach can also provide estimates of yield probability in target environments based on the understanding of G x E interactions. Such studies can help in reducing the number of sites/season required for field evaluation and thus increase the efficiency of the whole process of variety development.

13.7 Crop Model Limitations

Crop models are not able to give accurate projections because of inadequate understanding of natural processes and computer power limitation. As a result, the assessments of possible effects of climate changes, in particular, are based on estimations. Moreover, most models are not able to provide reliable projections of changes in climate variability on local scale or in frequency of exceptional events such as storms and droughts (Shewmake 2008).

General Circulation Models (GCMs) have so far not been able to produce reliable projections of changes in climate variability, such as alterations in the frequencies of drought and storms, even though these could significantly affect crop yields. GCMs do a reasonable job in simulating global values of surface air temperature and precipitation, but do poorly at the regional scale (Grotch 1988).

Furthermore, biological and agricultural models are reflections of systems for which the behavior of some components is not fully understood and differences between model output and real systems cannot be fully accounted for. Crop models are therefore not able to give accurate projections because of inadequate understanding of natural processes and computer power limitation. Again, methodology of model validation is still rudimentary.

Measured parameters also vary due to inherent soil heterogeneity over relatively small distances and to variations arising from the effects of husbandry practices on soil properties. Crop data reflect soil heterogeneity as well as variation in environmental factors over the growing period.

Model performance is limited to the quality of input data. It is common in cropping systems to have large volumes of data relating to the aboveground crop growth and development, but data relating to root growth and soil characteristics are generally not as extensive. Most simulation models require that meteorological data be reliable and complete. Finally, sampling errors also contribute to inaccuracies in the observed data.

An ultimate crop model would be one that physically and physiologically defines all relations between variables the model reproduces and universally real-world behavior. However, such a model cannot be developed because the biological system is too complex and many processes involved in the system are not fully understood (Jame and Cutforth 1996). Even if an ideal crop model could be produced, the collection of the highly precise system parameters and of the input data for the crop environment would be a formidable task in itself. Thus, the level of detail involved in a crop model is closely linked to the end use of the model and the precision required.

Even when a judicious choice is made, it is important that aspects of model limitations be borne in mind such that modeling studies are put in the proper perspective and successful applications are achieved. One of the issues still unresolved is the undue credit which the modelers are generally accused of taking from the data generators. This aspect is certainly worth looking into. There is always scope for improving existing models and other decision tools, and it will always remain a continuous process.

Development of a database is not as exciting as model development but equally important if not more so. There is a need to compile and reorganize existing data and to design, plan, and implement a global program with the sole objective of constructing a Minimum Data Set (MDS) for soil, crop, and weather variables. These characteristics can be integrated using geographical information systems (GIS) to form user-oriented decision support systems.

The development of new decision aids is taking place at such a fast pace that there is an immediate need to train a critical number of researchers, educators, administrators, and policy-makers so that the best possible benefit may be derived from them.

13.8 Some Common Crop Models

I. *DSSAT*

The Decision Support System for Agrotechnology Transfer (DSSAT) Version is a software application program that comprises crop simulation models for over 42 crops (as of Version 4.6).

For DSSAT to be functional, it is supported by database management programs for soil, weather, and crop management and experimental data and by utilities and application programs (Fig. 13.3). The crop simulation models simulate growth, development, and yield as a function of the soil–plant–atmosphere dynamics.

DSSAT and its crop simulation models have been used for many applications ranging from on-farm and precision management to regional assessments of the impact of climate variability and climate change. It has been in use for more than 20 years by researchers, educators, consultants, extension agents, growers, and policy- and decision-makers in over 100 countries worldwide.

The crop models require daily weather data, soil surface and profile information, and detailed crop management as input. Crop genetic information is defined in a crop species file that is provided by DSSAT and cultivar or variety information that should be provided by the user. Simulations are initiated either at planting or prior to planting through the simulation of a bare fallow period. These simulations are conducted at a

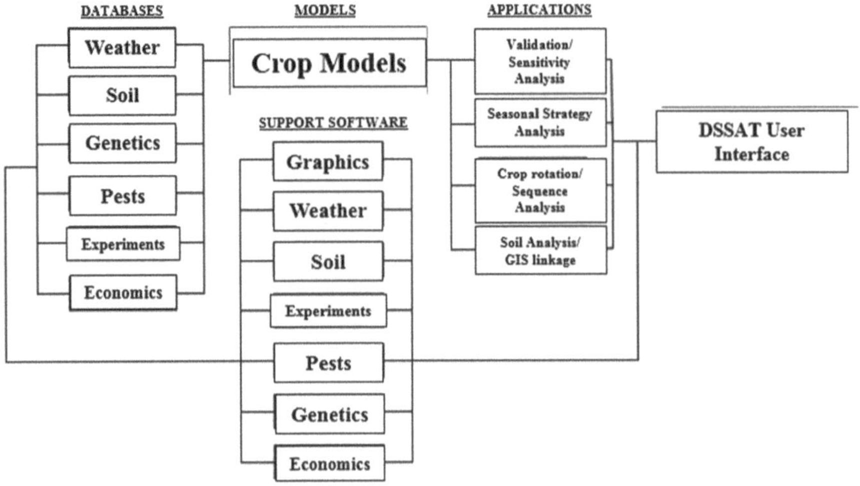

Fig. 13.3 Components of DSSAT

daily step and, in some cases, at an hourly time-step depending on the process and the crop model. At the end of the day, the plant and soil water, nitrogen, and carbon balances are updated, as well as the crop's vegetative and reproductive development stage.

For applications, DSSAT combines crop, soil, and weather databases with crop models and application programs to simulate multi-year outcomes of crop management strategies. DSSAT integrates the effects of soil, crop phenotype, weather, and management options and allows users to ask "what if" questions by conducting virtual simulation experiments on a desktop computer in minutes which would consume a significant part of an agronomist's career if conducted as real experiments.

DSSAT also provides for evaluation of crop model outputs with experimental data, thus allowing users to compare simulated outcomes with observed results. This is critical prior to any application of a crop model, especially if real-world decisions or recommendations are based on modeled results. Crop model evaluation is accomplished by inputting the user's minimum data, running the model, and comparing outputs with observed data. By simulating probable outcomes of crop management strategies, DSSAT offers users information with which to rapidly appraise new crops, products, and practices for adoption.

II. *WOFOST*

WOFOST (WOrld FOod STudies) is a simulation model for the quantitative analysis of the growth and production of annual field crops. It is a mechanistic model that explains crop growth on the basis of the underlying processes, such as photosynthesis, respiration, and how these processes are influenced by environmental conditions (Fig. 13.4).

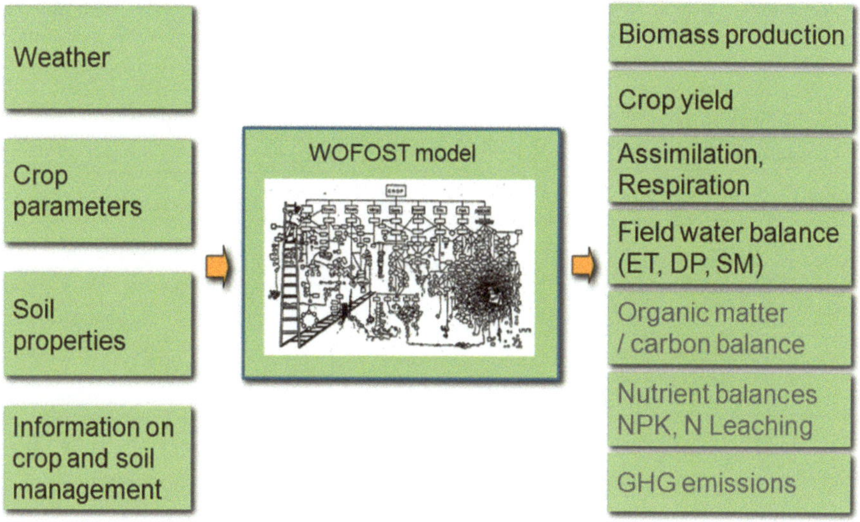

Fig. 13.4 Key inputs and outputs of WOFOST model

WOFOST can calculate attainable crop production, biomass, water use, etc. for a location given knowledge about soil type, crop type, weather data, and crop management factors (e.g., sowing date). WOFOST has been used by many researchers all over the world and has been applied for many crops over a large range of climatic and management conditions. Moreover, WOFOST is implemented in the Crop Growth Monitoring System which is used operationally to monitor arable crops in Europe and to make crop yield forecasts for the current growing season.

III. *InfoCrop*

InfoCrop, a generic crop model, simulates the effects of weather, soils, agronomic management (planting, nitrogen, residues, and irrigation), and major pests on crop growth, yield, soil carbon, nitrogen and water, and greenhouse gas emissions (Fig. 13.5). The model is user-friendly, requires easily available inputs, and is targeted to increase applications of crop models in research and development.

InfoCrop first considers the influence of weather, followed by soil factors, and pests, respectively, as has earlier been described by Rabbinge et al. (1994). The following processes are considered in the model:

- Crop growth and development: phenology, photosynthesis, partitioning, leaf area growth, storage organ numbers, source: sink balance, transpiration, uptake, allocation and redistribution of nitrogen
- Effects of water, nitrogen, temperature, flooding, and frost stresses on crop growth and development
- Crop–pest interactions: damage mechanisms of insects and diseases

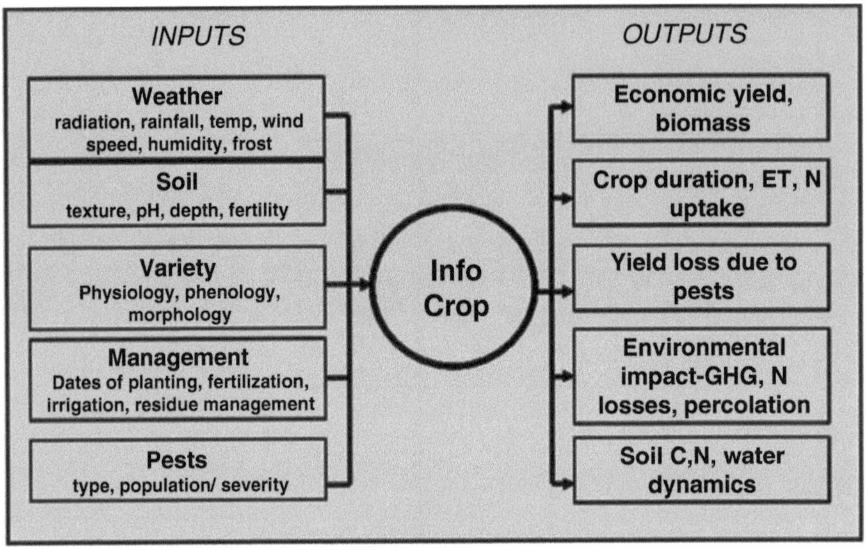

Fig. 13.5 Key inputs and outputs of InfoCrop model

- Soil water balance: root water uptake, interlayer movement, drainage, evaporation, runoff, ponding
- Soil nitrogen balance: mineralization, uptake, nitrification, volatilization, interlayer movement, denitrification, leaching
- Soil organic carbon dynamics: mineralization and immobilization
- Emissions of greenhouse gases: carbon dioxide (CO2), methane (CH4), nitrous oxide (N2O)

IV. *AquaCrop*

AquaCrop is the crop growth model developed by FAO to address food security and assess the effect of the environment and management on crop production. AquaCrop simulates the yield response of herbaceous crops to water and is particularly well suited to conditions in which water is a key limiting factor in crop production. The main components of the soil–plant–atmosphere continuum and the parameters driving phenology, canopy cover, transpiration, biomass production, and final yield in AquaCrop are shown in Fig. 13.6.

V. *APSIM*

The Agricultural Production Systems sIMulator (APSIM) is internationally recognized as a highly advanced simulator of agricultural systems. It contains a suite of modules which enable the simulation of systems that cover a range of plant, animal, soil, climate, and management interactions. APSIM is an undergoing continual development, with new capability added to regular releases of official versions. Its development and maintenance are underpinned by rigorous science and software

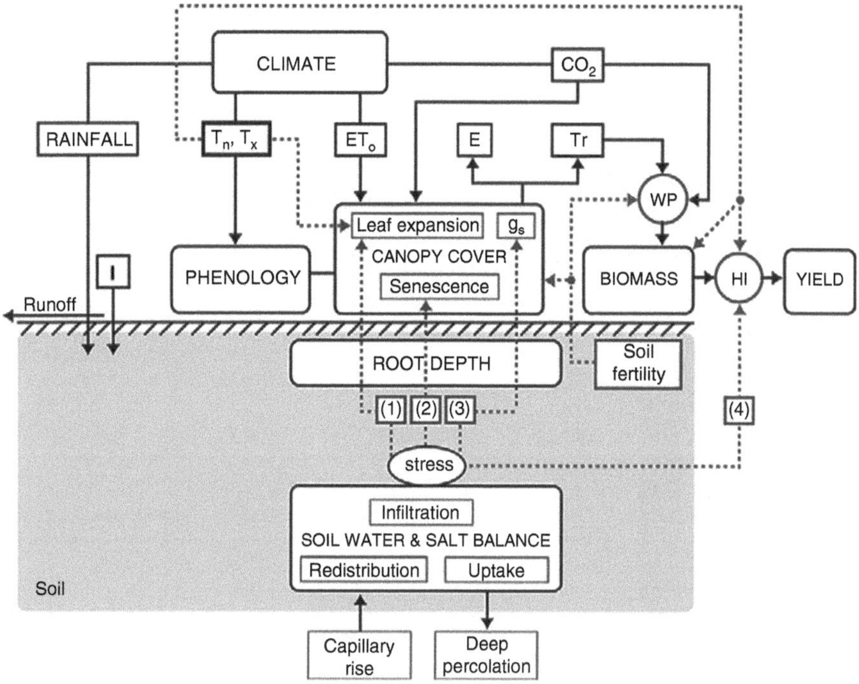

Fig. 13.6 Chart of AquaCrop indicating the main components of the soil–plant–atmosphere continuum and the parameters driving phenology, canopy cover, transpiration, biomass production, and final yield. *I* irrigation, T_n minimum air temperature, T_x Max air temperature, ET_o reference evapotranspiration, *E* soil evaporation, T_r canopy transpiration, g_s stomatal conductance, *P* water productivity, *HI* harvest index, CO_2 atmospheric carbon dioxide concentration; (1), (2), (3), (4), different water stress response functions. Continuous lines indicate direct links between variables and processes. Dashed lines indicate feedbacks

engineering standards. The APSIM Initiative has been established to promote the development and use of the science modules and infrastructure software of APSIM.

13.9 Conclusions

Models are means to capture, condense, and organize knowledge. A tested model can be a very effective scientific tool for tailor-made introduction of new production technologies, working out alternative crop production strategies, providing answers to the “what if” questions raised by technology adopters, identifying problems and prioritizing research, optimizing precious resources by reducing the number of field experiments, assisting in policy and strategy applications, environmental characterization and agroecological zoning, seeking new domains, and a very effective teaching aid.

References

Brockington, N. R. (1979). *Computer modeling in agriculture* (p. 156). New York: Clarendon Press.

Grotch, S. L. (1988). Regional inter comparisons of general circulation model predictions and historical climate data. Washington, DC: U.S. Department of Energy DOE/NBB-0084, 291 pp.

Jame, Y. W., & Cutforth, H. W. (1996). Crop growth models for decision support systems. *Canadian Journal of Plant Science, 76*(1), 9–19.

Kumar, R., & Chaeturvedi, S. (2009). Crop modeling: A tool for agricultural research.

Matthews, R. B., Kropff, M. J., Horie, T., & Bachelet, D. (1997). Simulating the impact of climate change on rice production in Asia and evaluating options for adaptation. *Agricultural Systems, 54*, 399–425.

Oteng-Darko, P., Kyei-Baffour, N., & Ofori, E. (2013a). Yield of rice as affected by transplanting dates and plant spacing under climate change simulations. *Wudpecker Journal of Agricultural Research, 2*(12), 055–063.

Oteng-Darko, P., Yeboah, S., Addy, S. N. T., Amponsah, S., & Owusu Danquah, E. (2013b). Crop modelling: A tool for agricultural research–A review. *E3 Journal of Agricultural Research and Development, 2*, 1–6.

Rabbinge, R., Van Diepen, C. A., Dijsselbloem, J., De Koning, G. J. H., Van Latesteijn, H. C., Woltjer, E., & Van Zijl, J. (1994). Ground for choices: A scenario study on perspectives for rural areas in the European community. In *The future of the land* (pp. 95–121). Wiley.

Shewmake, S. (2008). *Vulnerability and the impact of climate change in South Africa's Limpopo River Basin* (p. 804). International Food Policy Research Institute.

Singh, A. K. (1994). *Crop growth simulation models* (pp. 497–509). New Delhi: IASRI.

Xu, Q. (1995). A model for simulation of hybrid rice seed production. Applications of systems approach in plant breeding. SARP Research Proceedings, International Rice Research Institute. *PO Box, 933*, 97–108.

Glossary

A-B line The imaginary reference line set for each field for a tractor/sprayer guidance system to follow. There are different reference lines that can be set in a field to fit a particular geography or layout.

Accuracy (of GPS receivers) The measure of closeness of an object's actual (true) position to the position obtained with a GPS receiver. Accuracy levels are used to rate the quality of GPS receivers.

Aerial imagery Photos taken from airplanes, satellites, or UAVs to assist growers to determine variations within an area of interest such as a field.

Aerial photography Photos taken from airplanes to assist growers to determine variations within an area of interest such as a field.

Agriculture anomaly An agronomic (vegetation or soil) deviation or inconsistency in excess of "normal" variation from what one would expect to observe.

Algorithm A mathematical formula that may be used to control variable rate applications.

Application A practical use of computer software, an electronic system, or a concept.

Applications package Specialized computer programs and their associated documentation developed for practical usage. Ideally, applications packages allow a noncomputer specialist to use the computer without learning complex programming languages.

Archive The storage of historical records and data. When you have collected a year or two of data from your precision farming applications, you have started your own archive.

ASCII (American standard code for information interchange) A standard coding system used to represent alphanumeric characters within a computer. ASCII files enable the transfer of some data between different computers through the use of a common set of symbols.

Aspect Horizontal direction in which a slope faces (e.g., a SE-facing slope has an aspect of 135 degrees).

L. Ahmad, S. S. Mahdi, *Satellite Farming*,
https://doi.org/10.1007/978-3-030-03448-1

Attribute value A value or property that is a characteristic of a spatial element. For example, a specific symbol or color may represent 150–160 bushels/acre that is a value assigned to that attribute.

Attribute A numeric and/or text description of a spatial entity (e.g., address or owner's name for a parcel).

Auto-steer A GPS guidance system that steers agricultural equipment with centimeter accuracy. This level of accuracy requires real-time kinematic (RTK) correction of GPS signals. Auto-steer is an add-on component for equipment. It includes both the GPS system to receive and process the signals, software and hardware to allow the input of control maps, and the mechanical equipment to actually steer the tractor. Some new tractors are available "auto-steer ready."

Base map A simple map that shows the boundaries of a field or section and information about any unique feature (sinkholes or streams).

Base map The outline of your field with its proper coordinates is your base map. Data collected within the field by your yield monitor will be defined in location by the base map, which is a binary digital map.Base stationThe RTK-GPS receiver and radio that are placed in a stationary position, functioning as the correction source for roving tractor units in an area. These stations can be either portable or permanently installed systems, and their coverage can range from 5 to 10 miles depending on topographic conditions, antenna height, and radio-transmit power.

Benchmark (1). Used to define how comparisons are to be made between different computer software or systems according to specific requirements. (2). In surveying, a benchmark is the elevation at a specific point.

Bit An abbreviated term for binary digit, the smallest unit of computer data.

Block kriging A piecewise form of kriging based on grid cells.

Boom/section controller An electronic device that is capable of turning on/off sections of a chemical application boom manually or automatically in combination with GPS positioning and area mapping.

Buffer An area defined by the specified length extended around a point, line, or area.

Centroid The position at the center of a one- or two-dimensional (2D) entity such as a polygon.

Choropleth map A thematic map such as a yield image where quantitative spatial data is depicted through the use of shading or color variations of yield ranges.

Contour line A line used to represent the same value of an attribute (elevation or yield).

Contour map Yield map that combines dots of the same intensity/yield level by interpolating (or kriging).

Contour A line connecting a set of points, all of which have the same value. A contour line will show elevations of the same value.

Contouring Interpolation method used to distinguish between different levels of an attribute (elevation, fertility, yield).

Control segment The network of tracking stations that monitor and control GPS satellites.

Controller An electronic device used to change product application rates on the go, based on user directions or prescription application maps.

CORS (Continuously Operating Reference Station) (network) A network managed by the US Office of National Ocean Service (NOAA) to provide GNSS data consisting of carrier-phase measurements throughout the United States.

Crop scouting Precise assessments of pest pressure and crop performance that can be tied to a specific location for better interpretation.

Cross tabulation Comparison by location of attribute data in two or more map layers.

Data input The entry of information into a computer through the use of a keyboard, digitizer, and scanner or even entering data from already existing databases.

Data layer (in GIS) A layer of information on a GIS map. A map can have many layers to present different types of information. For example, the first layer of a map may be a satellite image of an area. The next layer may have only lines that represent roads or highways. The next layer may contain topographic information and so forth.

Data standardization The process of achieving agreement on common data definitions, representation, and structures to which all data layers must conform.

Database management system (DBMS) A collection of software for organizing the information in a database that might contain routines for data input, verification, storage, retrieval, and combination.

Database A collection of different pieces of georeferenced information (yield, soil type, fertility) that can be manipulated (layered) in a GIS model.

Database A logical collection of files managed as unit. A GIS database includes data about both the position and the attributes of geographic features.

DEM (digital elevation model) A digital representation of the elevation of locations on the land surface. A DEM is often used in reference to a set of elevation values representing the elevations at points in a rectangular grid on the earth's surface. Some definitions expand DEM to include any digital representation of the land surface, including TiMS or digital contours.

Differential correction Correction of the GPS signal to make it more accurate. An uncorrected signal will be accurate to about 50 yards. A corrected signal can be accurate to within 1–5 feet. Correction of a signal is done from a second GPS receiver/transmitter at a known fixed location. The signal is then transmitted to the tractor, combine, or other equipment that corrects the proper location through differential processing.

Differential global positioning system (DGPS) A method of using GPS which attains the position accuracy needed for precision farming through differential correction.

Digitizer A table or tablet which has the capability of digitally recording the relative position of a cursor which is moved over the area or line that you want to digitize or record.

DLG (digital line graph) A US geological survey digital map format used to distribute topographical maps in vector form. The digital files contain lists of the coordinate points that describe linear map features.

Edit The process of adding, deleting, and changing data/information on a computer.

Extrapolation A method or technique to extend data or inferences from a known location to another location for which the values are not known.

Feature A geographic component of the earth's surface that has both spatial and attribute data associated with it (e.g., field, well, waterway).

Field (1). A set of alphanumeric characters comprising a unit of information. (2). A location in a data record in which a unit of information is stored. For example, in your database, one of your crops may contain columns of data such as location #, crop type, variety, date of planting, etc. (all of which are fields). (3). A specific location on a person's farm that may be called "field # 10a."

Field prescriptions Applications of inputs at variable rates based on data obtained through yield monitors, crop scouting, remote sensing, and soil sampling.

Flow sensor A sensor that measures the amount of flow through an enclosure (tube, pipe, or housing) per unit of time.

Geocode A code associated with a spatial element which describes its location. An example would be a coordinate such as longitude or latitude.

Geographic data Data that contains information about the spatial location (position) and the attribute being monitored (yield, seed population, etc.). Also referred to as spatial data.

Geographic information system (GIS) A computer-based system used to input, store, retrieve, and analyze geographic data sets. The GIS is usually composed of map-like spatial representations called layers which contain information on a number of attributes such as elevation, land ownership and use, crop yield, and soil nutrient levels.

Geographic information systems (GIS) System of computer hardware, software, and procedures designed to support the compiling, storing, retrieving, analyzing, and displaying of spatially referenced data for addressing planning and management problems.

Georeference system A coordinate system keeping track of specific points on the earth's surface. Examples of such a system are the universal transverse mercator (UTM) system and the state plane coordinate system.

Georeferencing The process of adding geographic data to yield data or other field attributes either in real time (on the go) or by post-processing or the process of associating data points with specific locations on the earth's surface.

Geostationary satellite An orbital path of a satellite that is synchronized with the earth's orbit or space vehicles in an orbit which keeps them over the same location on the earth at all times.

Geostationary satellites Space vehicles in an orbit that keeps them over the same location on the earth at all times. Satellite-based differential correction signals are broadcast from this type of satellites. Others are maintained by NOAA to provide weather images every 30 min of the earth.

Global positioning system (GPS) A system using satellite signals (radio waves) to locate and track the position of a receiver/antenna on the earth. GPS is a technology that originated in the United States. It is currently maintained by the US

Government and is available to users worldwide free of charge. There are 30 satellites in the GPS constellation.

GLONASS (Global'naya Navigatsionnaya Sputnikovaya Sistema) Russian version of the American GPS satellite system (Global Navigation Satellite System). It is a radio-based satellite navigation system operated for the Russian government by the Russian space forces with a constellation of 24 operational satellites in 2010.

GNSS (global navigation satellite system) The standard generic term for satellite navigation systems that provide geo-spatial positioning with global coverage using time signals transmitted from satellites. The US GPS and the Russian GLONASS are the only two fully operational GNSS. Top-of-the-line GNSS receivers can communicate with both GPS and GLONASS satellites effectively doubling the available reference satellites at any given time.

GPS (global positioning system) A network of satellites controlled by the defense department that is designed to help ground-based units determine their current location in latitude and longitude coordinates. Note that the term "GPS" is frequently used incorrectly to identify precision farming. GPS is only one technology that is used in precision farming to assist you to return to an exact location to measure fertility, pests, and yield.

GPS antenna The device that receives satellite signals from space. On most handheld GPS devices, the antenna is integrated into the receiver device. For machine GPS systems, the antenna is typically an external device that can be mounted on top of the vehicle, away from the receiver.

Grid mapping Predetermined locations in a field where soil or plant samples may be obtained for analysis. The test information can be used for assessing fertility needs and determining approximate locations for varying fertilizer and lime applications.

Grid sampling The collection of samples from small, uniform-sized cells based on a systematic grid laid out across a field. Grid location in the field is used to develop a field map for the attribute measured.

Grid A data structure that uses rectangular units or grid cells arranged in rows and columns to represent an area like a field.

Ground control point An easily identifiable feature with a known location that is used to give a geographic reference to a point on a yield image.

Ground control points Stationary objects/areas on the earth's surface that provide georeferenced points in a remote sensing image/aerial photograph.

Ground reference data The field collection of data that is used in the interpretation of information gathered from other sources such as a yield image or a remotely sensed image. Also known as ground truth but the preferred terminology is ground reference.

Guided crop scouting Assessment and recording of crop anomaly and conditions on a site-specific basis using a backpack GPS receiver and handheld computer. The system allows the user to record growth stage/maturity, plant vigor, presence of disease, and weed and insect infestation.

Hard disk A large capacity, mechanical, magnetic, computer storage device that stores your programs and data.

Hardware The various physical components of an information processing system such as a computer, view screen, plotters, and printers.

Image classification Processing techniques which apply quantitative methods to the values in a digital yield or remotely sensed scene to group pixels with similar digital number values into feature classes or categories.

Input An overused term that applies to the process of entering data into a computer. Also used to describe the actual data that are to be entered.

Internet An international network comprised of many possible dispersed local and regional computer networks in which one can share information and resources. Developed originally for military and then academic use, it is now accessible through commercial online services to the general public.

Interpolation Mathematical procedure for estimating unknown values from neighboring known data.

Kriging (creeging) An interpolation technique for obtaining statistically unbiased estimates of spatial variation of known points such as surface elevations or yield measurements utilizing a set of control points.

Kriging An interpolation method for obtaining statistically unbiased estimates for field attributes (yield, nutrients, elevation) from a set of neighboring points.

Landsat (land satellite) A series of US satellites used to study the earth's surface using remote sensing techniques.

Lat/lon The latitude and longitude that specifically describe a position on the earth. Latitude is the north to south position. Longitude is the east to west position. Precise locations are described in degree, minutes, and seconds. The lat/lon of Purdue University is 86 degrees, 55 min, 05 s latitude and 40 degrees, 25 min, 50 s longitude.

Layer A logical separation of mapped information representing common data (e.g., roads, soils, yields, vegetative cover, and soil tests).

Legend A map section containing explanations of symbols, colors, and/or shades that signify various elements and data values on the map. A yield map will contain a listing of yield values and the color denoting a range of yields.

Light bar (in machine guidance) A device connected to a GPS receiver typically consisting of a row of LED lights to provide the tractor operator with a visual guide, day or night. The light bar does not automatically steer the tractor or machine; rather it aids the operator in driving on the imaginary reference line.

LIS (land information system) A system for describing data about land and its use, ownership, and development. LIS refers to all aspects of handling the data such as collection, storing, checking, merging, manipulating, analyzing, and displaying.

Locational reference Referencing data collected by yield monitor, sensor, or other methods and relating it to a specific spatial position.

Lookup table A reference table containing key attribute values that can be linked or related to data usually collected at a specific location. An example would be physical and chemical data relating to a soil-mapping unit.

Menu A list of options displayed by a computer data processing program, from which the user can select an action to be initiated. These choices are usually displayed in the form of alphanumeric text but may be as icons.

Merge To take two or more maps or data sets and combine them together into a single coherent map or database without redundant information.

Metadata A term used to describe information about data. Metadata usually includes information on data quality, currency, lineage, ownership, and feature classification.

Mosaic Process of assembling GIS database files for adjacent areas into a single file.

NAVSTAR (navigation by satellite timing and ranging) The US-based global navigation satellite system that was funded by taxpayers and controlled by the DoD.

NDVI image The normalized difference vegetation index (NDVI) is a simple graphical indicator that can be used to analyze remote sensing measurements and assess whether the target being observed contains live green vegetation or not.

Network (1). A group of linked computers that are able to share software, data, and various hardware devices such as printers. (2). A geometric or logical arrangement of nodes and interconnecting lines.

Noise Random variations or error in a data set. Also an unwanted sound coming from the combine.

Orthophotograph An aerial photograph that corrects distortion caused by tilt, curvature, and ground relief.

Output The product of a computer process and analysis that may be displayed on a computer screen or as a printed map or tables of values.

Pixel A term used in remote sensing referring to the fundamental unit of data collection which is an abbreviation for "picture element." A pixel is represented in a remotely sensed image as a rectangular cell in an array of data values and contains a data value that represents a measurement of some real-world feature.

Point sampling A method of grid sampling in which a sample is taken in a 10–30 foot radius at the center point of each grid location.

Polygon An area enclosed by a line describing spatial elements, such as a similar yields range, land use, or soil type.

Precision farming Using the best available technologies to tailor soil and crop management to fit the specific conditions found within an agricultural field or tract.

Precision farming Managing crop production inputs (seed, fertilizer, lime, pesticides, etc.) on a site-specific basis to increase profits, reduce waste, and maintain environmental quality.

Prescribed application The dispensing of a material or chemical into the field on a prescribed or predetermined basis. A prescription map is generated by an expert (grower and/or agronomist) based on the information about the field in use before an application. The prescription determines how much of something will be applied.

Prescription file A computer-generated GIS file that assigns a value to a given geographical area. For example, nitrogen application rate.

Raster format Format for storing GIS spatial data in which the data is stored in cells which are addressed by rows and columns of the cells.

Raster-to-vector conversion A process in which one converts an image such as a yield map of grid cells into a data set layer of lines and polygons.

Rate controller An electronic device that varies the amount of chemical/plant nutrient applied to a given area.

RDBMS (relational database management system) A database management software system that organizes data into a series of records that are stored in linked tables. This provides the ability to relate different records, fields, and tables and aids data access and data transformation.

Real-time correction Correction of a GPS signal by simultaneously transmitting the differential correction information to a mobile receiver.

Real-time kinetic (RTK) Procedure whereby carrier-phase corrections are transmitted in real time from a reference receiver to the user's receiver. Depending on local availability, RTK corrections can be delivered by radio modem from an on-site base station or a state's CORS network or even over the Internet using Wi-Fi.

Receiver (in GPS hardware) A computer-radio device that receives satellite information by way of radio waves to determine the position of its antenna relative to the earth's surface. The antenna can be integrated into the receiver or connected externally with a cable.

Registration A process where one can geometrically align maps or images to allow one to have corresponding cells or features. This allows one to relate information from one image to another, or a map to an image, such as registering a yield image to a soil map to determine if soils are influencing the yield response.

Remote sensing The act of detection and/or identification of an object, series of objects, or landscape without having the sensor in direct contact with the object. The most common forms include color and color-infrared aerial photography, satellite imaging, and radar sensing.

Remote sensing The act of monitoring an object without direct contact between the sensor and object.

Resolution A way of detecting variation. In remote sensing, one has spatial resolution (the variation caused by distance separating adjacent pixels), spectral resolution (the variation from the range of spectral responses covered by a wavelength band), and temporal resolution (the variation caused by time over the same location).

Satellite constellation A system of 24 satellites that is owned by the US Department of Defense (DoD) that can determine location to within inches. There are usually at least four of these satellites that are in view 24 h a day. The DoD can intentionally introduce error into the signal during national emergencies. This error called "selective availability" would allow an accuracy of approximately 50 yards without differential correction.

Satellite A communication vehicle orbiting the earth. Satellites typically provide a variety of information from weather data to television programming. Satellites send time-stamped signals to GPS receivers to determine the position on the earth.

Scale The ratio or fraction between the distance on a map, chart, or photograph and the corresponding distance on the ground. A topographic map has a scale of 1:24,000, meaning that 1 inch on the map equals 24,000 inches (2,000 feet) on the ground.

Semi-variance A measure of the difference between two neighboring data points (defined as 1/2 the square of the difference between two values).

Site-specific crop management (SSCM) The use of yield maps, grid sampling, and other precision tools to manage the variability of soil and crop parameters and aid decisions on production inputs.

Site-specific management The use of high-tech equipment to enable the farmer/producer to thoroughly assess field conditions and the applications of inputs to help obtain the best possible financial return.

Software The programs, procedures, algorithms (set of rules), and their associated documentation, for a computer system.

Soil map A map that indicates differences in soil properties (texture, fertility, organic matter, pH, etc.) within a field.

Spatial data Data pertaining to the location, shape, and relationship among geographical features.

Spatial data Data that contains information about the spatial location (position) and the attribute being monitored such as yield, soil properties, plant variables, seed population, etc. synonymous with geographic data.

Spatial resolution The size of the smallest unit that can be identified by a remote sensing observation.

Spatial variability Differences in observed attribute(s) that are noted between locations in a field.

Spectral resolution The ability of sensing system to differentiate between electromagnetic radiations of different wavelengths.

Speed sensors Sensors that measure the rotational speed of a shaft or the reflection of radio or sound waves off the ground to determine forward speed.

Thematic map A map related to a topic, theme, or subject. These maps emphasize a single topic such as yield, soil type, or land ownership.

Topologically integrated geographic encoding and referencing (tiger) file The nationwide digital database developed by the US Bureau of the Census. Tiger files contain street addresses and census boundaries with accompanying population statistics.

Turn-key system A reference to hardware and/or software systems, meaning that they are ready to be used immediately and are designed, provided at a cost and supported by a commercial group.

Unmanned aerial vehicles (UAVs) An unmanned aerial vehicle (UAV), commonly known as a drone and also referred by several other names, is an aircraft without a human pilot aboard. The flight of UAVs may be controlled either autonomously by onboard computers or by the remote control of a pilot on the ground or in another vehicle. In agriculture, UAVs are typically used to survey crops. The available two types of UAVs – fixed wing and rotary wing – are both equipped with cameras and are guided by GPS. They can travel along a fixed flight path or be controlled remotely.

Variable rate application (VRA) Adjustment of the amount of crop input such as seed, fertilizer, lime, or pesticides to match conditions (yield potential) in a field.

Variable rate technology (VRT) System of sensors, controllers, and agricultural machinery used to perform variable rate applications of crop production inputs.

Variable rate technology Instrumentation such as a variable rate controller for varying the rates of application of fertilizer, pesticides, and seed as one travels across a field.

Vector format A format for storing and displaying GIS spatial data that is stored as points, lines, or areas to create a map object. By using a nearly continuous coordinate system, vector data can be more accurately georeferenced than raster data.

Waveband A remote sensing term used to describe a contiguous range of wavelengths of electromagnetic energy. Visible wavelengths (seen by the human eye) range from 400 to 700 nanometers. Near-infrared (NIR) wavelengths are at 700–2600 nanometers.

Wireless communication Data transfer and voice communications using radio frequencies or infrared light.

Yield map A map that indicates differences in crop yield within a field. Data is usually collected by a yield monitor on a combine over 1–3 s intervals.

Yield maps A representation of crop yields collected on the go by a harvester equipped with an instantaneous yield monitor. Each location/site (pixel) in a field is assigned a specific crop yield value.

Yield monitor Electronic device that continuously measures and records crop yields and moisture on the go.

Yield monitoring Regular intervals where a harvested weight has been obtained along with a GPS reading. A display of the weights every 1–3 s is translated to bushels/acre or yield providing a yield map. Moisture of the grain is obtained at the same time.

Zone management The information-based division of large areas into smaller areas for site-specific management applications.

MIX
Papier | Fördert gute Waldnutzung
FSC® C083411

Kolibri 360 GmbH
Ferdinand-Jühlke-Straße 7
99095 Erfurt, Deutschland
produktsicherheit@kolibri360.de